PATTERNS IN THE
BALANCE OF NATURE

THEORETICAL AND EXPERIMENTAL BIOLOGY

An International Series of Monographs

CONSULTING EDITOR

J. F. Danielli

State University of New York at Buffalo, Buffalo, New York, U.S.A.

Patterns in the Balance of Nature

AND RELATED PROBLEMS IN QUANTITATIVE ECOLOGY

C. B. WILLIAMS

1964

ACADEMIC PRESS

LONDON and NEW YORK

ACADEMIC PRESS INC. (LONDON) LTD
BERKELEY SQUARE HOUSE
BERKELEY SQUARE
LONDON, W.1

U.S. Edition published by

ACADEMIC PRESS INC.
111 FIFTH AVENUE
NEW YORK 3, NEW YORK

Copyright © 1964 by Academic Press Inc. (London) Ltd

Second printing 1968

Library of Congress Catalog Card Number: 64–14226

Re-printed Photolitho by
Page Bros. (Norwich) Ltd.

PREFACE

THE object of this book is to draw attention to a large number of problems in statistical ecology, and particularly in population balance, which have in common the property of being expressible as frequency distributions, and so have a mathematical pattern.

In the past, population studies have been chiefly concerned with the position of one species or of a very small number in association. In experimental studies the association and the environment have usually been so simplified that it is doubtful if the conclusions are always applicable to the infinitely complex conditions in the field.

Here the central theme is the pattern in which all the species in an association are balanced, irrespective of the particular position of one or two selected species. Year after year the pattern may remain the same while particular species are constantly changing in position. The approach is that of synecology and not aut-ecology. It is an attempt to emphasize the pattern of the whole mosaic, which has an identity and characteristics of its own, and in which the same result could be produced by many thousands of re-arrangements of the same units.

The survey is wide rather than detailed, and in most cases final conclusions are not reached, and often not attempted. There can be no doubt however that some of the concepts—such as that of a measurable diversity, which is a property only of the population and not of any of the units from which it is made—are valuable tools in the study of population ecology.

By concentrating on the pattern of balance, instead of the unit concerned, it is possible to see relations between ecological problems which otherwise appear unrelated. From the mathematical form of the pattern a start has been made to find past conditions of competition, survival, variation and evolution which might have resulted in such a pattern. Much more can be done in this way in the future.

These introductory remarks are made to enable the reader to see what was the object in view: they are not to forestall or disarm criticism. As a very young lady of my acquaintance once wrote from school to her mother "these are reasons—not excuses".

January 1964   C. B. WILLIAMS

TO MY WIFE

IN GRATITUDE FOR HER WISDOM

IN THE ART OF LIVING

CONTENTS

Chapter 1

INTRODUCTION

ANY animal or plant population, in the wild state, consists basically of a very large number of "individuals"; although in some cases, particularly in the plant world, even this simple unit is difficult to define. These individuals are classified by naturalists (who believe that they are thereby expressing real relationships and not merely convenient figments of their own imagination) into "species", the species into "genera", the genera into "families", and so on till we come to the two great kingdoms of "Animals" and "Plants".

The total number of living individuals existing at any moment is almost inconceivable, and certainly incalculable. There may be 1000 million bacteria in a single gram of soil; 100,000 diatoms in a litre of sea water; or 400 million insects and 700 million mites in the surface soil of an acre of forest land, without counting the inhabitants of the trees above. I have estimated the number of insects in the world at any moment to be of the order of 10^{18} or 10^{19}; or between 1 and 10 million million million (see p. 101).

The number of different species of animals already known and described for the whole world is over 1 million, of which the majority are insects; and new ones are being discovered every day. The total number of existing species of insects is quite possibly as many as 3 million, which would imply that not much more than one-quarter are already known. On the other hand, bird students recognize about 25,000 species, and consider that this is probably about 90 per cent of the true number. The discovery of a new bird is a much rarer event than that of a new insect. The number of species of flowering plants so far described is about 200,000, and this is undoubtedly well below the real number.

The number of individuals in different species varies within very wide limits. We know, for example, that the St Kilda Wren (*Troglodites troglodites hirtense*, Seebohm) has today probably only about 250 pairs, while on the other hand a single swarm of locusts may contain over 10,000 million individuals.

Both very small and very large numbers within a species may bring dangers. In bisexual animals low numbers result in a difficulty in finding a mate, hence very rare animals are usually also very localized. On the other hand, large numbers bring dangers of increase or concentration of enemies, or of epidemic diseases; and also dangers of starvation, particularly when the food supply itself fluctuates in abundance, either seasonally or over longer periods.

The number of individuals within a species is changing from moment to moment and from season to season, and the rapidity and extent of these changes is largely determined by the interactions of the birth rate, the length of life and the death rate before maturity. The elephant produces one young at a time, after a pre-mature life of 30 years or more, and man runs it close for slow breeding and slow change of numbers. In both these cases the rate of change in total population numbers is still further reduced by the over-lapping of several generations.

An insect, on the other hand, frequently produces a thousand eggs or more, and may complete its life cycle in a few weeks; and there is seldom any overlap in generations. A bacterium may double in size, and divide into two (hence doubling the local population) in less than half an hour.

A single flowering plant may live for many years, sometimes over a thou-sand, and may produce tens of thousands of seeds in a year, while a single fruiting body of a fungus may produce millions of spores in a few hours. Plants still further complicate problems of multiplication by vegetative reproduction.

In order that the number of individuals in a species should remain stable the death rate before maturity must balance the birth rate. So the fecundity of a species is an indication of the dangers of its early life. Life prolonged after reproduction is finished—with the possible exception of social organiza-tions as in ants, bees and man—is of no value to the species, and may even become a danger.

Amid this complexity the individuals of each species have to struggle for survival against the physical properties of their immediate environment, such as weather, changing seasons, and soil or water conditions; and also in competition for food, and against the danger of being eaten. This competition is with all other species of living creatures in their neighbourhood, and at times even with individuals of their own species, as, for example, in the choice of a mate, with only here and there the helping hand of symbiosis or commensalism. Relative success or failure in their struggles determines whether the number of individuals in the species is on the up-grade or falling.

As a result of this complex and endless interplay each species establishes temporarily an uneasy balance of numbers among all the others. The pattern of the relative abundance of all the species in a mixed community (and all wild communities are mixed) is thus a synthesis of all the competition and co-operation, and all the difficulties and facilities, that have surrounded all the species of the community in the recent past; and every minute the posi-tion of any one species is changing in relation to the others.

The pattern of relative abundance is thus an expression of the momentary balance which has been set up among all the species of the association, and it is important to find out whether, as time passes, the fundamental pattern changes, or if the species move in their relative abundance within a more or less stable pattern. This is the approach to the problem of the "balance of nature" from the point of view of quantitative synecology.

A certain amount of study, both biological and mathematical—with the biological work nearly all, unfortunately, under relatively simple and stabilized laboratory conditions—has been carried out on the "balance" between a host animal and its parasites, or between two or more species competing for space; but what happens in the complex natural conditions in the field has been almost completely neglected.

It is, of course, impossible to study all the individuals of all species over a large area, and so our material in the field has first to be limited in area and confined to certain definite groups, such as, for example, the birds of the British Isles, or of some piece of woodland; the insects in an acre of land; or the plants of some particular ecological association.

Even among these smaller populations some sampling method will probably be necessary, and new difficulties then arise as to which properties of the sample are also properties of the population sampled, and which are not. If a random sample of individuals is taken from a mixed wild population of animals containing a large number of species, there appears to be a mathematical order in the relative abundance of the different species represented. In general, more species are represented by one individual than by two, more by two than by three, and so on. The frequency distribution of the number of species with different numbers of individuals is of the "hollow curve" type. Since order cannot be made out of chaos by the mere process of sampling, we must infer that there is some related order in the relative abundance of the species in the population sampled.

Just as the number of individuals differs greatly from species to species, so, on a much smaller scale, the number of species differs from genus to genus. The new species constantly being discovered either add to the number in already known genera, or require the definition of new genera to contain them. Thus the number of described genera is steadily increasing. The genera and species that we know form a random sample of those that exist. If a classification in almost any group of animals or plants is studied statistically, it will be found that there are nearly always more genera with one species than with two, more with two than with three, and so on. The frequency distribution is not unlike that found in a random sample of individuals when classed into species, and has been noticed by various biologists, including Willis, Yule, Chamberlin and Small. The same relation appears to exist also between genera and families, and between families and higher groups.

If certain hosts are examined and the number of parasites on each is recorded, it not infrequently appears that, when the average number of parasites per host is low, the number of hosts with one parasite is greater than the number with two, there are more with two than with three, and so on. There is evidence of this with fleas on rats, and with lice in human beings. So again we find a somewhat similar mathematical pattern.

It is the main object of the present study to bring forward evidence from various sources to illustrate these curiously related problems; to discuss certain of the mathematical series that have been suggested at various times to fit the observed data; and especially to see what light the structure of these samples can throw on the pattern of frequency distribution in the population sampled, and so on the form of balance which has been assumed under natural field conditions.

If we can find a mathematical model that closely fits the observed data, we will be nearer to an understanding of our problem, for at such a stage we could see what follows mathematically from the theory and so devise further tests and experiments. Also it might be possible later to find what combination of previous conditions would bring about such a frequency distribution, and to see if these can be justified on biological grounds. Thus we could throw light on the mechanism of population balance.

Arising out of these central problems are a number of other questions of interest to ecologists. For example the theories lead to the possibility of measuring the "diversity", or the richness or poverty of variety in an animal or plant community, quite apart from the absolute abundance of species or individuals. This is dealt with more fully in Chapter 7.

An extension of the analysis to the relation between species, genera and higher groups in classification, throws light on many problems of classification (Chapter 6), and on the extent of competition between species considered to be sufficiently closely related in origin to be classified in the same genus (Chapter 9).

In the plant world a study of the distribution of species in small quadrats (Chapter 4) leads to a better understanding of the influence of the number and size of the quadrats on the interpretation of field data. It shows that the mathematical conception of diversity contains many of the older ideas on measurable characteristics of a population in a form independent of the size and number of the quadrats.

A further extension to much larger areas (Chapter 5) demonstrates the more rapid increase of species with area when the areas (unlike the smaller quadrats) include within and between themselves a diversity of environment, and in the limit a diversity of evolutionary origin.

It is also suggested that the conception of diversity can be used to measure the relation between different faunas or floras (Chapter 5), and finally that low specific diversity is characteristic of a population in an environment where the physical conditions are severe or specialized. In an easier physical environment, such as in the tropics or sub-tropics, the chief determining factor of the balance is biological competition, and the resulting diversity is high.

Chapter 2

THE MATHEMATICAL APPROACH

PROBLEMS OF SAMPLING

ONE of the main difficulties in studying patterns of abundance in animals and plants is to obtain reliable information from wild natural populations about the frequency of species represented by 1, 2, 3, etc., individuals; of genera with 1, 2, 3, etc., species; of hosts with 1, 2, or more parasites; or in general of groups with 1, 2, 3, etc., units.

Such information could be most accurately obtained by taking all the units in the particular population and identifying each into its proper group; this would, in fact, be a complete census of the population. This ideal is, however, very seldom possible. The alternatives are either to make an estimate of the numbers of individuals over the whole population or to get a more accurate count in a randomly selected sample. Both these methods introduce errors, and the object must be to adopt the technique which, in the particular circumstances, reduces the errors to a minimum and if possible allows the residual error to be measured.

In field estimates the greatest errors are human, and it is very important that all numbers should be checked by several observers. Estimates of the size of a flock of birds, or the number of animals in a field can often differ, between observers, by 200 to 300 per cent. Human errors in estimation of numbers in the field are likely to be biased in one direction, and also there is a greater liability to "mistakes", such as wrong identification, which cannot be corrected by later checking.

It follows that the most usual practice is to take an actual sample from the field, which can be sorted at leisure, and can in many cases be kept for future reference. The sample must be random, by which is meant that the technique of sampling must not produce any bias in the factor, e.g. abundance, which one wishes to study. As human selections are liable to this error, it is always desirable to have some mechanical method, such as some form of trap. Large samples are theoretically better than smaller ones; but often one has to choose between less accurate large or more accurate small samples. More information, especially about the size of errors, can be obtained by several small samples, which can later be combined to one larger one.

To get a true cross-section of the pattern of a population the sample should be as instantaneous as possible. In some cases, such as plankton sampling with a net, this can be achieved. In field observations of birds, and in some forms of trapping, the sampling period may extend to several hours or days; in other cases several weeks or months may be necessary before a sufficiently large sample can be obtained. Even when trapping lasts only a few hours the period of activity of the different members of the population may be so limited that the chance of capture may vary with the length of activity during the trapping period.

In many insect studies the trapping is confined to the active winged adult stage, and this has to be taken into consideration in studying populations at different periods of the year. The relative length of life of different species, and even of the two sexes of a species may affect the results of different methods of sampling. Traps that depend on the attraction of insects to a light or bait are less easy to interpret than those, such as a suction trap, which are independent of any special sense of the animal, except perhaps its ability to escape.

In plant populations the problem of movement does not occur, but there is the added difficulty of defining an individual unit, as discussed on p. 65.

According to the particular problem, or the necessities of the situation, samples may be taken in two different ways: either by a random selection of complete groups or by a selection of units which are later classified into groups. In the first it must be possible to select groups within which all the units are contained. For example, the distribution of fleas on rats can be studied by catching a number of rats (groups) and finding out how many fleas (units) there are on each. It would not be possible to take a random selection of fleas and then to see to which rat they belonged. An increase in the size of the sample adds to the number of groups, but does not add any units to the groups already counted.

In the second case there is a random selection of units which are later sorted into their proper groups. For example, a number of insects (units) may be collected in a random manner by some form of trap, and then sorted into species (groups). In this case an increase in the size of the sample will add new units to groups already present, and probably also add a small number of new groups to those already present in the first sample.

This second method is most frequently used in studying the relative abundance of species as it is never possible to collect all the individuals of a species.

Having obtained the sample, one sees if there is any evidence of a pattern or mathematical order in the frequency distribution. If there is such an order, the next step is to see if it can be represented—closely if not exactly—by any known mathematical law, such as a normal distribution, or some series of which the properties are known.

The next stage is to enquire what part of any such law or order reflects a similar or related condition in the population sampled, and what is a property of the sample only, or some function of the sample size. It is only the former which is of biological importance.

As an example of the way in which the process of sampling can produce misleading results, we can take the case of a population consisting of 1500 species (units) classified into 500 genera (groups). In this population the average number of species per genus is three. However, in any random sample of species the smaller the sample the smaller the average number of species per genus. In the limiting case of a sample of one species being selected, the average number of species per genus can only be one; and this average steadily increases with sample size till the sample contains the whole population. In fact, only by taking a random sample of whole genera (sampling by groups) can the average number of units per group remain constant (see p. 256).

Thus the average number of units per group, when the sample is randomized by units, is a function of the size of the sample. This is a matter of mathematics and not of biology, yet in many cases it has led biologists astray.

It must also be realized that the process of sampling, even if perfectly random, may alter the apparent pattern of the sample from that of the population, and the pattern of the best possible sample is not necessarily a miniature reproduction of the original. For example, if a population consisted of a number of groups each with an equal number of units, then in any sample the groups would not be of equal size; in fact, their frequency distribution would be in the form of a Poisson Series.

THE ZERO TERM; GROUPS WITH NO UNITS

In studying any animal or plant association consisting of a large number of species (groups) with different numbers of individuals (units), a species which was not represented by even a single individual would not normally be considered as a part of the population. If, however, only samples are available for study some of these may contain species not represented in others, and so some will have no individuals of species which are present in others although these species presumably belong to the population sampled. If surveys are taken in successive years the same situation arises, but the size of the zero term would depend on the number of years' study. Over a long period there might also be slow changes in the environment which would have its effect on the survival of different species.

Thus mathematical formulae which require an accurate knowledge of the zero term are difficult to use except where the sample is large enough to contain the great majority of the species present in the population.

If, on the other hand, we are sampling by groups—as, for example, when trapping a number of rats and counting the number of fleas on each—the number of zero groups (rats without fleas) is a real and accurately measurable part of each sample. When the average infestation is low, the number of hosts without parasites may be a high proportion of the total. If it were possible in such a population to make a random sample of fleas, and then find out to which rat they belonged, there would be no indication of the number of zero groups.

THE ZERO TERM IN THE DISTRIBUTION OF PARASITES ON HOSTS

When a large number of some host animal or plant are examined for the presence of parasites there may or may not be, according to the general level of infection, a certain number of hosts without any parasites. It is important when studying the place of these in any mathematical pattern to consider their possible origin.

If, in a laboratory experiment, a number of hosts are exposed to infection by parasites, the experiment is usually designed so that each of the hosts is equally likely to be found by the parasite. In this case the zero term is of the same significance as the numbers of infected hosts and is part of the statistical distribution. If, however, a large sample is obtained in the field, it cannot be assumed that each of the hosts had an equal chance of being selected or rejected by the parasite. Small differences in the position of the host might alter its chances of being discovered. The distribution of the parasites might be irregular, or only cover part of the range of the host. Thus the zero term, according to the conditions of sampling, might include hosts which were never liable to infection: those which, being in conditions of possible infection, were either overlooked or rejected by the searching parasite; or often a mixture of the two. It is therefore necessary to use extreme care in the mathematical interpretation of the number of hosts without parasites in field studies.

TRANSFORMATION OF SCALE OF DATA FROM ARITHMETIC TO GEOMETRIC OR SQUARE-ROOT CLASSES

If there is evidence that any measurable character under investigation is not varying arithmetically (which may be inferred from a skew distribution on an arithmetic scale), it is often useful to transfer the data to geometric (logarithmic), square root, or other type of class interval to see if a symmetrical result is obtained. Not only is such a type of distribution easier to analyse, but the transformation required often gives information about the processes involved in the variation.

To do this it is desirable that the divisions between the classes should be equally spaced on the new log or root scale. It is not desirable, for example, to take as a geometrical series the values 1,   2 + 3,   4 + 5 + 6 + 7,   8 to 15, . . . as, although each class contains twice the number of integers of the previous one, the dividing lines at 0.5, 1.5, 3.5, . . . are not equally spaced on a logarithmic scale.

Preston (1948), wishing to transform to a geometric scale data on the number of animals with different numbers of individuals per species, used classes in what he called "octaves"* in which the dividing points on the logarithmic scale were at 0.0, 0.30, 0.60, 0.90, etc., or at 1, 2, 4, 8, 16, etc., on the arithmetic scale; each class has double the range of the previous one. Since the only possible values in his problem were integers, it is necessary to split the observed numbers at each integer which

---

* The word "octave" is borrowed from the musical world in which each octave corresponds to a doubling of the frequency of vibration of the note, but its actual derivation is from the eight notes into which the scale is divided by harmony. But as no conception of eight subdivisions comes into this type of problem the word introduces a false association. If a word is needed, "doublet" would be a better expression.

is a dividing line between the class above and the class below (see Table 1).

It should be noted here that no logarithmic scale can take into consideration true zero values (the log of which is minus infinity), but its use envisages classes below unity with divisions at $\bar{1}.70$, $\bar{1}.40$, $\bar{1}.10$, $\bar{2}.80$, and so with fractional limits on the arithmetic scale at 0.5, 0.25, 0.125, 0.0625 . . . Some of the zeros in biological data undoubtedly should come into these fractional classes, as, when a group occurs on an average in one out of ten samples, its true place is in the 1/10 class, although in nine out of ten it appears as a zero (see p. 12).

TABLE 1. *Upper and lower limits (on the arithmetic scale) of the first fourteen classes on a geometric (logarithmic) scale at intervals of $\times$ 2 and $\times$ 3*

Class	$\times$ 2			$\times$ 3
	Half of	+ all +	half of	
I	—	—	1	1
II	1	—	2	2–4
III	2	3	4	5–13
IV	4	5–7	8	14–40
V	8	9–15	16	41–121
VI	16	16–31	32	122–364
VII	32	33–63	64	365–1093
VIII	64	65–127	128	1094–3280
IX	128	129–255	256	3281–9841
X	256	257–511	512	9842–29,524
XI	512	513–1023	1024	29,525–88,573
XII	1024	1025–2047	2048	88,574–265,720
XIII	2048	2049–4095	4096	265,721–797,161
XIV	4096	4097–8191	8192	797,162–2,391,484

A slightly coarser geometric scale, depending on a class increase of $\times$ 3 instead of $\times$ 2 (0.48 on the log scale instead of 0.30), has the advantage of requiring no splitting of integer values. In 1953 I proposed a scale with the division line below unity at $\bar{1}.70$ on the log scale and further divisions at 0.18, 0.65, 1.13 . . ., which correspond to 0.5, 1.5, 4.5, 13.5 . . . on the arithmetic scale. All class divisions come exactly half-way between two integers; the integers and fractions coming into each successive class for the two series are shown in Table 1. Thus on the $\times$ 2 scale, fourteen classes cover the range from 1 to 8192, but on the $\times$ 3 scale the same number of classes range to over 2 million.

Slightly larger classes can be obtained using a scale of $\times$ 5 or $\times$ 7. If the first division is taken at 0.5 on the arithmetic scale, then, so long as an odd number is used as a multiplier, all divisions will come between the integers and no splitting will be necessary. For example, in the $\times$ 5 scale the successive classes will contain $1 - 2$, $3 - 12$, $13 - 62$ . . .

A still coarser scale, not infrequently used for very large ranges, is based on $\times$ 10, and has dividing lines at 1, 10, 100, 1000, etc. At each of these numbers the observed values have to be split equally between the upper and the lower classes.

In certain population studies, particularly of a theoretical nature, it is sometimes desirable to use classes based on a "× e" scale, where "e" is the base of the natural logarithms and is equal to 2.718 approximately. Although somewhat difficult for a biologist to appreciate, it bears a definite relation to the formulae of several frequency distributions, including the normal and the log-normal. It also bears a relation to the measurement of the standard deviation and of diversity.

THE SQUARE-ROOT TRANSFORMATION

If it is required to test a square-root transformation (which can include true zero values), once again it is desirable that the divisions between the classes are at equal intervals. Two simple methods, both of which, however, require the splitting of integer values, are as follows:

(1) to have the points of class divisions at 0, 1, 2, 3 . . . on the square-root scale corresponding to 0, 1, 4, 9, 16 . . . on the arithmetic scale;

(2) to have the centre points of the classes at these integers and the dividing points half-way between at 0.5, 1.5, 2.5, 3.5 . . . corresponding to 0.25, 2.25, 6.25, 12.25 . . . on the arithmetic scale.

TABLE 2. *Upper and lower limits (on the arithmetic scale) of the first fourteen classes on the square-root scale; with the divisions at the integers, and with the integers at the centres of the divisions*

Class	Divisions at integers on arithmetic scale			Centres of classes at integers on arithmetic scale		
	Half of +	all +	half of	$\frac{1}{4}$ of +	all +	$\frac{3}{4}$ of
I	—	—	0	—	—	0
II	0	—	1	0	1	2
III	1	2, 3	4	2	3, 4, 5	6
IV	4	5–8	9	6	7–11	12
V	9	10–15	16	12	13–19	20
VI	16	17–24	25	20	21–29	30
VII	25	26–35	36	30	31–41	42
VIII	36	37–48	49	42	43–55	56
IX	49	50–63	64	56	57–71	72
X	64	65–80	81	72	73–89	90
XI	81	82–99	100	90	91–109	110
XII	100	101–120	121	110	111–131	132
XIII	121	122–143	144	132	133–155	156
XIV	144	145–168	169	156	157–181	182

The contents of these classes by the two methods on the original arithmetic scale are shown in Table 2. It will be seen that fourteen classes only take the upper limit up to 169 or 182 respectively on the original scale. To reach 1 million there would have to be 1000 classes. If it were necessary to

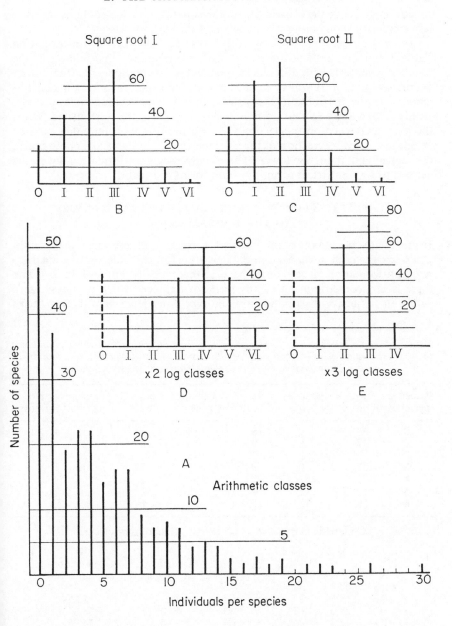

FiG. 1. The frequency distribution of copepods parasitic on mussels, showing the original data on an arithmetic scale of abundance (A); and also transformed to the square-root scale by two methods (B and C); and to the log scale in classes differing by × 2 (D) and × 3 (E). (See Table 3.)

cover a large range the class limits on the square-root scale could be taken at larger intervals combining two or more of the smaller classes.

As an example of the working of these various methods of transformation Table 3 and Fig. 1 show the results with the frequency distribution of the number of mussels with different numbers of a copepod parasite (for further details see p. 230). It will be seen that with the original figures on an arithmetic scale, the distribution is very skew, with the peak at the zero level, and a steady falling away so that no mussel has more than thirty parasites. When the log transformation (neglecting the zero term) is used on either × 2 or × 3, there is overcorrection of the asymmetry so that the peak is now nearer the upper end than the lower. The square-root transformation gives the nearest to a symmetrical form, but both forms are slightly truncate.

THE ZERO TERM WHEN THE SCALE IS TRANSFORMED TO THE LOGARITHMIC

In the square-root, or any other root, transformation all zero values in the arithmetic scale remain unchanged in the new transformed scale. With a logarithmic transformation all zero terms become theoretically "minus infinity", and there is no geometric mean of any series of numbers which includes a zero value. In frequency distributions of the type considered here the zero term

TABLE 3. *Examples of the transformation of data from an arithmetic scale of abundance to geometric and square-root scales, as illustrated by the frequency distribution of parasitic copepods in mussels in Poole Harbour, Dorset, England*

Parasites per host	No. of hosts	Parasites per host	No. of hosts	Parasites per host	No. of hosts	Parasites per host	No. of hosts
0	47	7	16	14	4	22	1
1	37	8	9	15	2	23	1
2	19	9	7	16	1	26	1
3	22	10	8	17	2	30	1
4	22	11	7	18	1		
5	14	12	4	19	2	Total	
6	16	13	5	20	1	250	

	Geometric classes			Square-root classes	
	× 2	× 3		A. Divisions at integers	B. Class centres at integers
Zero	47	47			
I	18.5	37	I	23.5	35.25
II	28.0	63	II	42.0	63.0
III	42.5	86	III	70.5	74.75
IV	61.5	17	IV	69.5	54.0
V	42.0	—	V	34.0	18.0
VI	10.5	—	VI	8.5	4.75
			VII	2.0	0.25

can be considered as including those with theoretically fractional values below unity as mentioned above.

As it is not usually possible to sort out the zero term in this way an approximation, which has considerable practical value particularly in obtaining a mean, can be made by adding a small number x to the number of units in each class and then considering the log $(n + x)$ instead of log n.

In 1937 I suggested the use of log $(n + 1)$, which when $n = 0$ gives the log $1 = 0$, and when $n = 1$, the log $(1 + 1)$ is 0.30. This addition has the effect of making the earlier terms, up to about 10, very close to the square-root scale; from values of n from 10 to 50 there is a gradual change over to nearer the log scale; and at values of n over 50 there is little difference between the values of log n and log $(n + 1)$ (see Williams, 1937). Gaddum (1945), and Kleczkowski (1949) have discussed other values for the small addition.

The arithmetic mean of the values of log n is, of course, the log of the geometric mean of the values of n. The arithmetic mean of the values of log $(n + 1)$ is the log of the geometric mean of the values of $n + 1$. This transformation should only be used when zero values are present.

Another method of approach to the zero term, particularly when accumulated totals are required, was first suggested by my son J. L. Williams and is discussed briefly by Aitchison and Brown (1957, p. 104). It is based on the assumption that the particular two-parameter variate can only be manifested at the discrete integers 0, 1, 2, 3 . . . If the basic variation is geometric, these integers can be considered as the geometric means of the class intervals. If we define the class boundaries as lying at a_1 (between 0 and 1) and a_2 (between 1 and 2), etc., then all the values between 0 and a_1 will appear as 0; all the values between a_1 and a_2 will appear as 1, and so on.

TABLE 4. *Lower and upper limits of the arithmetic integer classes when the scale of variation is geometric*

Integer as geometric mean of class	Upper and lower limits of class	
	Arithmetic	Logarithmic scale
0	all below 0.637	all below $\bar{1}$.804
1	0.637–1.571	$\bar{1}$.804–(0.00)–0.196
2	1.571–2.546	0.196–(0.301)–0.406
3	2.546–3.534	0.406–(0.477)–0.548
4	3.534–4.527	0.548–(0.602)–0.656
5	4.527–5.522	0.656–(0.699)–0.742

It follows that $a_1a_2 = 1^2$, $a_2a_3 = 2^2$, etc., and Aitchison and Brown state " . . . it can be shown for consistency that a_1 is uniquely determined as equal to $2/\pi$" which equals 0.6366, and the log of this is $\bar{1}$.8039. Thus for purposes of summation the zero term can be considered as the sum of all terms up to 0.6366 on the arithmetic scale, or up to $\bar{1}$.8039 on the log scale.

Table 4 shows on both the arithmetic and the logarithmic scales the limits of the earlier integer classes according to this theory. It will be seen that they rapidly approach to the half-way line between the two successive integers, so that at the higher values the arithmetic and the geometric means between two successive units are practically indistinguishable.

If one adopts the × 3 class intervals already discussed, to reduce the number of classes, then the zero (fractional) classes would all sum up to a log value of $\bar{1}.804$. The class containing the first integer would have its upper limit at 0.196; and the class including the integers 2, 3 and 4 would have its upper limit, on the log scale at 0.656. As this is the log of 4.527, the geometric mean is already within 0.6 per cent of the arithmetic mean.

THE LOGARITHMIC TRANSFORMATION AND THE STANDARD DEVIATION

When a frequency distribution is symmetrical on an arithmetic scale it can be defined by the mean plus or minus the standard deviation.

When a frequency distribution is skew on an arithmetic scale, but can be brought to approximate symmetry by a log transformation, as in the case of a log-normal distribution, the distribution on the log scale can be defined by the mean log plus or minus the standard deviation on the log scale. For example, a log-normal distribution might have a mean at 2.00 and an S.D. of 0.30, which would usually be written as 2.00 ± 0.30.

It must be recollected, however, that if these figures are transferred back to an arithmetic scale, not only must the logs be replaced by their anti-logs, but the + and − of the log scale become × and ÷ on the arithmetic scale. Thus the suggested log-normal 2.00 ± 0.30 when transformed back to an arithmetic scale has the mean at 100 and a standard deviation of 2. But the limits of the standard deviations above and below the mean are not 100 plus or minus 2, but 100 multiplied or divided by 2. Thus the definition of this distribution on the arithmetic scale must be written $100 \times / \div 2$. The expression of the S.D. as ± is not only meaningless but misleading.

UNITS, GROUPS AND MOMENTS

The basic approach to frequency distributions in this study is that of a number of units (say individuals) belonging to a number of groups (say species) and the classes are those groups with different numbers of units. In the practical field side of the work all numbers of groups and of units are integers and positive. Fractional values can only occur when averages are used. Negative terms cannot occur in any consideration.

Mathematically the distribution must be a discontinuous series in which n_r is the number of groups with r units. In such a distribution Σn would be the total number of groups, and also the zero moments of the series. $\Sigma n r$ would be the total number of units, and also the first moments of the series. The second moments $\Sigma n r^2$ has no simple biological equivalent, but is used in the measurement of diversity, and will be discussed later (p. 149).

Thus in problems in which we usually start with a knowledge of the total groups and units, it is important to have methods of analysis which can be based on the moments of a mathematical distribution.

THE NUMBER OF PARAMETERS OR CONSTANTS

In testing the fit of field data to theoretical distributions it is important to remember that the more constants or parameters there are in the formula, the closer will be the fit without any increase in the logical proof of applicability. In fact, if one had as many parameters as observations, a perfect fit could always be obtained without any progress in understanding.

The hyperbolic series has only a single parameter, which can be the first term, the number of groups with only one unit. The log series can be calculated from two parameters, e.g. the number of groups and the number of units. The normal distribution, the log-normal and the negative binomial each require three parameters. In the latter case the first term n_1 is often taken as one of the constants, and it is important to remember that when this has been done, the remarkable resemblance between the observed and calculated number of groups with one unit cannot be taken as a convincing proof of the identity of the two distributions.

SOME FREQUENCY DISTRIBUTIONS

The number of possible mathematical frequency distributions is infinite; certain distributions have, however, in the course of time been found to have biological applications, and some are based on logical assumptions that could apply to the conditions of population balance.

The properties of the logarithmic series are discussed in more detail in Appendix A (p. 307). Here we will give only a brief summary of some of the others, and a diagrammatic scheme of their relations to each other. The principal points to notice are whether the distributions are continuous or discontinuous (i.e. integer values only); how many parameters are involved; if the series is convergent or divergent (i.e. if the sum of units is finite or infinite); and if negative values are included, as these have no meaning in biological frequencies although they may be necessary stages in an analysis.

Taking as a base the *positive binomial*, its simple formula is $(p + q)^k$ where p and q are the chances of two alternative happenings, and k is the number of times the situation is repeated. The series formed by the expansion is p^k; $k/2(p^{-k1} q)$, etc. If $p + q$ add up to 1 we get a probability series, in which case the sum of all the terms is also 1. If an actual series is considered the sum is finite, and the classes are integral if k is an integer.

If $p = q$ (i.e. there are equal chances of one alternative event happening or not happening), we get the familiar *normal distribution* so frequently used in statistics. It is theoretically a continuous curve, but can be applied, with some restrictions, to discrete integer series. As a probability series, with p and q adding up to 1, it has only a single parameter (the standard deviation), but for an actual frequency distribution of units in groups two constants are necessary.

If p is much larger than q (i.e. only a small chance of one of the alternatives happening) and if k is large, we get the *Poisson series*, which gives the frequency distribution of a rare occurrence in a large number of repetitions.

If the formula of the positive binomial is changed to $(p — q)^{-k}$ it forms the series known as the *negative binomial*, which is mathematically of great interest and has been found to be applicable to a number of biological problems. It is, however, difficult for a biologist to visualize. It is an integer series, and to fit to observed data three constants are necessary.

When in the negative binomial, the value of k is very small we get the *logarithmic series*, which is in another form the expansion of $— \log_e (1 — x)$. This is also an integer series, convergent, and requires only two constants to fit to observed data.

An extreme form of the log series results in the Hyperbolic or a harmonic series having the form 1, 1/2, 1/3, etc. It is divergent, and only sums to infinity. To fit to known data only requires one constant.

The hyperbolic series is also a special case of the inverse power series $x^k y = $ a constant, when k is equal to 1. Both this and the hyperbolic series have been suggested as applying to biological data, but the inverse power series cannot be applied to any finite number of groups and units unless the value of k is over 2. This rules out the hyperbola.

Since it has become more and more apparent that in most biological problems involving frequency distributions the variation of the number of units per group is geometric and not arithmetic, it is necessary to consider mathematical formulae based on this assumption. Thus we get the *log – normal* distribution, which differs from the normal, from our point of view, only in the fact that the successive class limits are at geometric and not arithmetic intervals. On an arithmetic scale of classes data in such a form will give a skew curve, which, however, becomes symmetrical and "normal" when the classes are transformed.

Just as the normal distribution leads to a log-normal so the Poisson series can lead to a *log – Poisson*. Unfortunately the theoretical development of this is so recent that it has not been possible to give it full consideration in this work (Cassie, 1962).

The diagram below shows the relation of the different series.

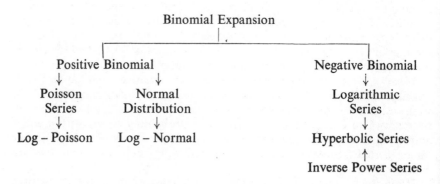

Binomial Expansion

Positive Binomial   Negative Binomial

Poisson   Normal   Logarithmic
Series   Distribution   Series

Log – Poisson   Log – Normal   Hyperbolic Series

Inverse Power Series

Chapter 3

PROBLEMS OF INDIVIDUALS AND OF SPECIES, CHIEFLY ZOOLOGICAL

THE quantitative approach to the study of the numbers of individuals, and the relative abundance of species, in animal and plant populations is of comparatively recent growth. It is, however, of interest to note the existence over seven centuries ago of a statistically minded naturalist.

In Volume 3, p. 422, of the second edition of Silvestre de Sacy's "Chresomathie Arabie" (1827) there is an extract from an old Arabic manuscript, "Book of the Marvels of Nature and the Singularities of Created Things", written about the first half of the thirteenth century by Mohamed Kaziwini (or Qaswini). This states that Khalif de Samarkande one night collected round the light a measure called Macouc of moths (*papillon* in the French translation) and on dividing them he counted seventy-three different species (*espèces* in the translation, but probably better expressed as "kinds"). According to Dr Griffiths, of Edinburgh University, the measure Macouc or Makkuk properly means a drinking cup, and so probably might contain between ½ and 1 litre.

Khalif de Samarkande seems to have been born before his time!

Coming to more modern times, in 1928 Stanley Garthside, in an unpublished thesis of the University of Minnesota, discussed the relative abundance of species in some collections of insects made by sticky traps and sweep nets in a forest area in Itaska Park, Western Minnesota. He identified 399 species from the 5186 individuals caught in the traps, and 488 from the 5665 individuals in the sweep nets. They belonged to seven different orders, but chiefly Diptera and Hymenoptera. He commented that "one of the outstanding facts is the extremely large number of species which occur in very small numbers . . . especially in the Hymenoptera where 80 per cent of the species occurred less than six times". He also showed that in none of the orders was the proportion of species with one to five individuals less than 48 per cent. His data are summarized in the first two columns of Table 5, and are shown diagrammatically in Fig. 2.

His results were commented on by Chapman (1931) and by Graham (1933), the latter adding two new sets of data, from Ohio, which are also shown in Table 5 and Fig. 2.

Garthside made no attempt to provide a mathematical interpretation, but repeated "we are forced to conclude that, despite the large number of individuals that characterize some species, the great bulk of species occur in relatively small numbers".

TABLE 5. *Frequency distribution of numbers of species of insects with different numbers of individuals in samples captured by Garthside in Minnesota, U.S.A., and by Graham in Ohio, U.S.A.*

| | Number of species | | | |
| | Garthside | | Graham | |
Individ. per sp.	Sticky trap	Sweep net	A	B
1	171	159	126	71
2–5	118	159	67	29
6–10	39	43	12	8
11–20	33	34	7	6
21–50	16	24	6	4
over 50	22	29	3	8
Total species	399	448	221	126
Total individuals	5186	5665	not given	

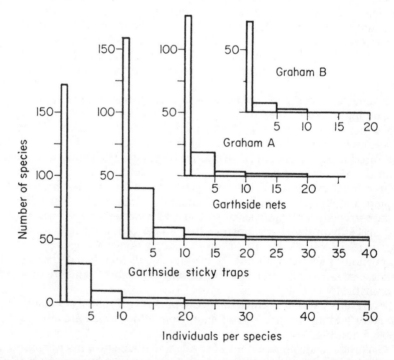

FIG. 2. Frequency distributions of species of insects with different numbers of individuals in field samples taken in Minnesota, U.S.A., by Garthside, and in Ohio by Graham (see Table 5).

From the diagrams in Fig. 2 it will be seen that the frequency distributions give, on an arithmetic scale of abundance, the typical hollow curve.

In 1942 Corbet pointed out that in a large collection of butterflies he had made in Malaya, the 9031 individuals included 620 species. Of these, 118 were each represented by only a single individual, seventy-four were represented by two individuals, forty-four by three individuals, and so on as shown in Table 6. From a knowledge of the total butterfly fauna of the area, he also estimated that 304 other species, known to occur in the district, were not represented in his collections, thus giving a possible zero term to the series.

TABLE 6. *The number of species of Malayan butterflies with different numbers of individuals in collections made by A. S. Corbet (1942). The zero term (*), which represents the species not caught, is the difference between the total number of species known to occur, and those represented in the collection*

Individ. per sp.	No. of sp.	Sub-totals	Individ. per sp.	No. of sp.	Sub-totals	Individ. per sp.	No. of sp.	Sub-totals	Individ. per sp.	No. of sp.	Sub-totals
0	304*										
1	118	(118)	21	11		41	1		64	1	
2	74		22	5		42	2		66	1	
3	44		23	3		43	1		68	1	
4	24	(142)	24	3		44	1		70	1	
5	29		25	5		45	4		71	4	
6	22		26	4		46	2		76	1	
7	20		27	8		47	—		84	1	
8	19		28	3		48	2		89	1	
9	20		29	3		49	1		92	1	
10	15		30	2		50	3		93	1	
11	12		31	5		51	1		100	1	
12	14		32	4		52	2		105	1	
13	6	(157)	33	7		52	1		108	1	
14	12		34	4		54	4		119	1	(53)
15	6		35	5		55	1		141	1	
16	9		36	3		56	5		147	1	
17	9		37	3		57	—		194	1	(3)
18	6		38	3		58	2				
19	10		39	3		59	1				
20	10		40	1	(147)	60	2				

Total individuals = 9031; total species = 620

For the first time an attempt was made to find a mathematical expression that would fit the data, and Corbet remarked that it resembled the "hollow curve of Willis" (1922) (see p. 119). Further, he says that "if the values given are converted to their respective logarithms, it is found that between $n = 1$ and $n = 24$ the plot closely approximates to a straight line; and the relation between S, the number of species, and n, the number of individuals per species, is expressed by the equation $S = C/n^m$, where m and C are constants." The straight-line relation that he indicates is when $\log S + m \log n = \text{constant}$ ($\log C$), which represents a hyperbola if $m = 1$.

Unfortunately, as Corbet himself pointed out, his sample was not a truly random one, as more attention was paid to the catching of a rare species than a common one. When about twenty-five individuals of any one species had been captured little effort was made to obtain others. Thus the sample was biased in the direction of too many species with small numbers, and too small numbers of individuals for the commoner species.

Corbet sent his preliminary results to R. A. Fisher, then working at Rothamsted, where I already had available the results of 4 years' continuous trapping of Lepidoptera in a light trap, including over 16,000 moths identified to about 240 species. The great advantage of these data was that the trap did not cease to collect any further specimens of a species just because it was abundant so that very large numbers of some of the commoner species had been captured.

As a result of an examination of these new data the joint paper of Fisher *et al.* was published (1943) giving a new suggestion, based on theoretical mathematical reasoning, for the interpretation of the frequency distribution —the logarithmic series. In relation to this theory, we reviewed Corbet's data on Malayan butterflies, and my own on Lepidoptera and other insects captured in light traps at Rothamsted Experimental Station, Harpenden, about 25 miles north of London.

Fisher's suggestion was that the frequency distribution of the species represented by 1, 2, 3, etc., individuals was, not the hyperbola which is divergent, but the logarithmic series, which is an integer series with a finite sum, and takes the form

$$n_1, \quad n_1 x/2, \quad n_1 x^2/3, \quad n_1 x^3/4, \ldots n_1 x^{n-1}/n$$

for the number of groups with 1, 2, 3, etc., units per group, when x is a constant less than unity (see Appendix A, p. 307).

If we divide n_1 by x and call the result α we get an even simpler expression of the series as follows:

$$\alpha x, \quad \alpha x^2/2, \quad \alpha x^3/3, \quad \alpha x^4/4, \ldots$$

This series is convergent and the sum of all the groups (in this case the species) $S = \alpha(- \log_e \overline{1 - x})$, and the total number of units (individuals) $N = \alpha x/(1 - x)$. There are only two parameters and only one possible series for any combination of a given number of groups and units. If we know these two we can find α and x, and so can calculate the logarithmic series that fits the data.

The constant "α" has been found to be a measure of the diversity of the population, and is low when the number of species is low in relation to the number of individuals, and high when the number of species is high. Fisher has shown that all random samples of individuals from a population arranged in a logarithmic series are also logarithmic series, and have the same value of α as the population sampled, but a lower value of x. Thus the structure of the sample indicates the diversity of the population sampled. For a fuller discussion of diversity see Chapter 7.

Figure 126 (Appendix A, p. 311) shows the rate of increase of the number of groups (species) in relation to an increase in the number of units (individuals) in the logarithmic series. The relation is given by $S = \alpha \log_e (1 + N/\alpha)$. It will be seen that for any particular value of the diversity, that is to say in samples of increasing size taken from the same population, there is at first for small samples a slightly curved relation between S and log N, which gradually straightens out to an almost straight line. The earlier portion is when the "1" in the above formula is significant in relation to N/α; the straight portion is where it is relatively so small that it can be neglected. The formula then approximates to $S = \alpha \log_e N/\alpha$. Thus for larger samples the increase in number of species resulting from an increase in the number of individuals from N_1 to N_2 is given by:

$$S_2 - S_1 = \alpha (\log_e N_2 - \log_e N_1) = \alpha \log_e N_2/N_1.$$

From this it follows that if the size of the sample is multiplied by e ($= 2.718$) $n_2/n_1 = e$ and therefore the increase in number of species $S_2 - S_1 = \alpha$.

When a logarithmic series was calculated to fit Corbet's uncorrected data of 9031 individuals in 620 species it gave a value of $x = .997$ and an expected n_1 of 135 species, which is considerably more than the observed 118. The calculated and observed values for $n_1 - n_4$ were 145 and 142, and for $n_5 - n_{13}$ were 145 and 157. Thus the estimated number of species is too high for the rarest, becoming too low for the commoner species. (For fuller calculations see Fisher et al., 1943, p. 43.) In 1944 I made an attempt to calculate values of x and n_1 using only the more reliable figures for (1) the number of species with less than twenty-five individuals, (2) the number of individuals in these species, and (3) the total number of species in the collection. It was not possible to get consistent results from these, but, by assuming that the assiduity of collecting fell off a little below the level of twenty-five individuals per species, which Corbet was willing to allow, consistent results were obtained with $n_1 = 131$ and $x = .991$.

It was not considered that this was a proof of the applicability of the logarithmic series to the data, but it showed that the differences between this particular theory and the observed numbers could be caused by departures from random sampling which might easily have occurred. (For fuller details see Williams 1944, p. 15.)

The first tests for the logarithmic series with unbiased data were made with Macro-Lepidoptera captured in a light trap at Rothamsted Experimental Station, about 25 miles north of London. Three examples of these results are shown in Fig. 3. They are (A) the total figures for 4 years' trapping from 1933 to 1938, (B) the average values per year, and (C) the average for catches covering one-eighth of a year, obtained by taking in each year eight sets of every 8th night. The diagrams also show, as dotted lines, the numbers expected on the assumption of the logarithmic series.

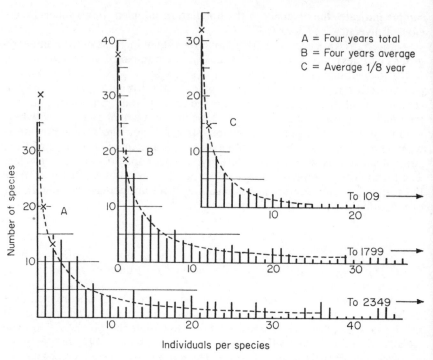

FIG. 3. Frequency distribution of species of Macro-Lepidoptera, with different numbers of individuals, from catches in a light trap at Rothamsted Experimental Station, in S.E. England. The dotted line shows values estimated from the logarithmic series distribution.

It will be seen that the distribution for the total of 4 years is more irregular than the other two sets which are averages. Also in this the observed n_1 and n_2 (species with one and two individuals) are below the calculated, and n_3 — n_6 above. In the other two sets n_1 is larger than calculated. The fit on the whole was close and justified further tests, several of which are given later in this chapter.

In 1948 Preston suggested that the frequency distribution of species with different numbers of individuals might be better represented by a log-normal distribution than by the log series. This distribution (see p. 15), which is strictly a continuous curve and not an integer series, is the normal, Gaussian, distribution when the dimensions of the variate (individuals per species) are expressed in geometric and not arithmetic classes. Preston suggested that small samples from such a population would be a truncate log-normal and, when shown on an arithmetic scale, would closely resemble a logarithmic series.

It is therefore important to see first how these two alternative suggestions can be distinguished, particularly with smaller samples; next to see if the data at our disposal are sufficiently critical to support one rather than the other; and lastly to see what are the implications of accepting one or the other. There is also the possibility that in different cases one gives a better interpretation, or that some new alternative is better than either.

In the first place, attention must be drawn to the fact that the log-normal distribution has three constants or parameters, while the logarithmic series has only two. Where therefore both formulae are possible, the former might give a better fit without necessarily being a better explanation. For a given number of individuals and species in a population or sample, there is only one possible log series, but many possible log normals. The form of the log normal is determined by the number of species (the area of the curve) and the standard deviation, which is a measure of the spread. The position of the median is determined by the number of individuals in relation to the number of species.

In a log-normal distribution, if the classes of number of individuals per species are grouped geometrically we get a normal or truncate normal according to the size of the sample. If the distribution is truncate it may be possible to make an estimate of the number of species below the class including those with one individual. These are the species which are not included in the particular sample, but which might be included in another, so they are a measure of the zero term. This estimate is more accurate in larger samples where more than half the species are represented, as then the median and peak of the distribution are available.

It is therefore interesting to see what happens if data in a logarithmic series are treated in the same way. Table 7 shows the results, on a × 3 basis

TABLE 7. *Theoretical frequency distribution of species, in × 3 classes, in samples of different sizes from a population arranged in a logarithmic series with* $\alpha = 100$. *For other values of* α *the numbers of species must be multiplied by 100*

Size of sample				Number of species in class			
Indi-viduals	Species	Average individ. per sp.	x	I (1)	II (2–4)	III (5–13)	IV (14–40)
43	37	1.16	0.3	30	5.60	0.065	—
100	69	1.44	0.5	50	18.23	1.09	—
233	120	1.94	0.7	70	41.94	8.32	—
900	230	3.91	0.9	90	81.28	47.91	10.94
1900	300	6.34	0.95	95	94.07	72.61	33.50
2173	306	6.95	0.956	95.6	95.6	—	—
3746	360	10.5	0.974	97.4	99.77	88.56	58.65
9900	461	21.47	0.990	99.0	105.36	101.00	86.05
19,900	531	37.48	0.995	99.5	106.95	105.27	97.25
99,900	691	144.6	0.999	99.9	108.04	108.50	107.18
Inf.*	Inf.	Inf.	1.000	100	108.33	109.68	109.84

* Hyperbolic or harmonic series.

(see p. 9), of taking samples of different sizes from an imaginary population arranged in a logarithmic series with a diversity of 100. The different samples have different average numbers of individuals per species and so different values of x. The expected numbers of species in each of the first four classes is given.

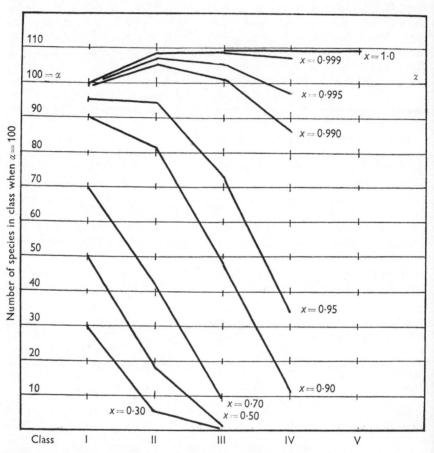

FIG. 4. Theoretical frequency distribution of species, in × 3 log classes, in samples of different sizes taken from a population arranged in a log series with α = 100 (see Table 7).

It will be seen most clearly from Fig. 4 that for the smaller samples, with values of x below 0.90, the number of species in each succeeding class is smaller than that in the preceding one. Thus the number of species with one individual is larger than the total number with two, three or four individuals. When x is 0.95 these first two classes are approximately equal, but then a rapid fall starts. With $x = 0.99$ the number in Class III is greater than in either

I or II. But however large the sample, x cannot exceed 1.0, and so none of the classes after I can exceed this by more than 10 per cent. In the theoretical case of an infinite sample $x = 1$, the curve becomes an hyperbola; the number of species in successive geometric classes then is asymptotic to just less than 10 per cent above the number in Class I (n_1).

LEPIDOPTERA IN A LIGHT TRAP AT ROTHAMSTED EXPERIMENTAL STATION IN 1935

In Table 8 and Fig. 5 are shown analyses of captures of Macro-Lepidoptera in a light trap about 25 miles north of London in the year 1935, including 6814 individuals of 197 species. The table shows the number of species at each level of individuals per species; the calculated values up to n_{13} on the

TABLE 8. *Frequency distribution of abundance of Lepidoptera captured in a light trap at Rothamsted Experimental Station during 1935*

Individ. per sp.	Species obs.	log series	Individ. per sp.	Species	Individ. per sp.	Species	Individ. per sp.	Species
1	37	38.0	15	2	39	1	87	1
2	22	18.9	16	2	40	3	88	1
3	12	12.5	17	4	42	2	105	1
4	12	9.3	18	2	48	2	115	1
5	11	7.4	20	4	51	1	131	1
6	11	6.1	21	4	52	1	139	1
7	6	5.2	22	1	53	1	173	1
8	4	4.5	23	1	58	1	200	1
9	3	4.0	25	1	61	1	223	1
10	5	3.6	28	2	64	2	232	1
11	2	3.3	29	2	69	1	294	1
12	4	3.0	33	2	73	1	323	1
13	2	2.7	34	2	75	1	603	1
14	3	—	38	1	83	1	1799	1

Total individuals = 6814; total species = 197

Class	In × 3 classes Species obs.	log series	Accumulated percentage Without zero	With zero assumed = 30
0	—	—	—	13.2
I	37	38	18.8	29.5
II	46	40.7	42.1	49.8
III	48	39.8	66.5	70.9
IV	37		85.3	87.2
V	19		94.9	95.6
VI	8		98.9	99.1
VII	1		99.5	99.6
VIII	1		100	100

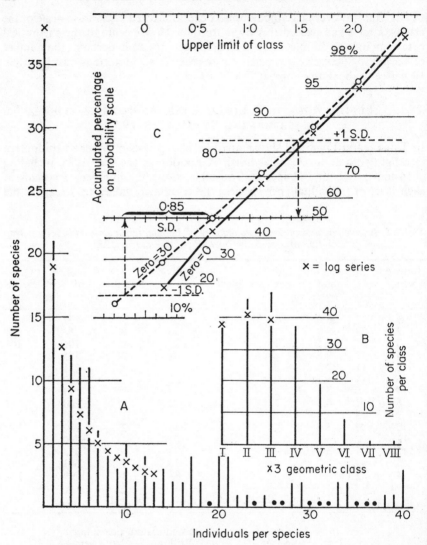

Fig. 5. Frequency distribution of species of Macro-Lepidoptera with different numbers of individuals, caught in a light trap at Rothamsted Experimental Station in 1935.

assumption of a log-series distribution; and the number of species in successive × 3 classes of abundance. There is also shown the accumulated percentage of the total species in these geometric groups first on the assumption of no zero value, and secondly on the assumption of a zero value of 30, which is the approximate number inferred from Fig. 5B on the basis of symmetry.

The figure shows the frequency distributions diagrammatically, and also the two sets of accumulated percentages plotted on a probability scale against log number of individuals per species.

It will be seen that the log series gives a very close estimate for n_1 (and hence for group I) and for n_3, but underestimates n_2 and n_4, and in the $\times$ 3 groups underestimates II and III. On the probability diagram there is a close fit to a straight line, particularly on the assumption of a zero of thirty species. This indicates a close fit to a log-normal distribution with a median, on the log scale, at 0.65 and a standard deviation of 0.85. On the arithmetic scale this becomes a median of 4.5 individuals per species with a S.D. of $\times / \div$ 7.

When a negative binomial was tested against the same data, and assuming as before a zero term of 30, the calculated n_1 was 20 instead of the observed 37. Even when the assumed zero term was raised to 100, the calculated n_1 was only 26. Thus it appears that the negative binomial, as applied to this set of data, implies a zero term which has little or no biological meaning.

LEPIDOPTERA IN A LIGHT TRAP AT ROTHAMSTED
IN DIFFERENT YEARS

In the course of 8 years' trapping at Rothamsted with several different traps 90,000 Macro-Lepidoptera were captured and identified to just under 350 species. Of this total nearly 33,000, of 285 species, were in a single trap (A) which was in continuous use in one position for two series of 4 consecutive years, 1933–37 and 1946–50. From these results, and particularly from trap A, it is possible to get a series of samples of increasing size from below 500 individuals, taken on every 8th day in a year, to 87,000 in four different traps over a total of 16 trap-years.

Some of the results are given in Table 9 and Fig. 6, with, for each sample, the total number of individuals and species, and also the species with different numbers of individuals classified in a $\times$ 3 geometric scale. Further data will be found in Williams (1953a).

TABLE 9. *Number of species of Lepidoptera (in $\times$ 3 classes of abundance) captured in light traps at Rothamsted Experimental Station in periods ranging from 1/8 year to 16 trap-years*

Individ.	Species	I (1)	II (2–4)	III (5–13)	IV (14–40)	V (41–121)	VI (122–364)	VII (365–1093)	VIII (1094–3280)	IX (3281–9841)	X (9842–29,524)
	A. One-eighth year; average 1933–6; trap A.										
492*	88	36	27	17	7	0.6	0.2	0.1	—	—	—
	B. 1 year; average of eight; trap A.										
3754*	175	42	47	40	26	15	3.6	1.0	0.3	—	—
	C. Total 4 years; average of two periods; trap A.										
16,065*	244	39	42	52	49	31	23	6	1	0.5	—
	D. Total 8 years; 1933–6 and 1946–9; trap A.										
32,853	285	38	41	48	67	43	28	14	5	1	—
	E. 16 trap-years; 4 traps. Total.										
87,400	346	37	47	38	56	61	61	33	9	3	1
	F. July 1949 only; average of six traps.										
1230	120	37	34	26	16	7	0.2	—	—	—	—

* Geometric mean.

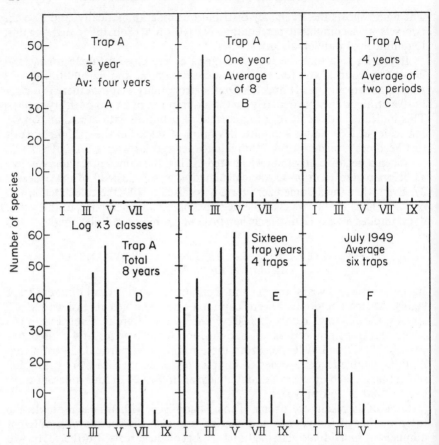

FIG. 6. Distribution of species of Macro-Lepidoptera with different numbers of individuals, captured in light traps at Rothamsted Experimental Station over periods of from 1/8 year to 16 trap-years with four traps (see Table 9). The scale of abundance is × 3 log classes.

It will be seen that the peak frequency in this log-scale distribution is in Class I (one individual per species) for the samples taken every 8th night, with an average of 492 individuals and eighty-eight species. It is also in this class for the single-month average (Fig. 6F) with 1230 individuals in 120 species, but here the following fall is less rapid. As the period of sampling is increased to 1 year, giving average catches of 3754 moths in 175 species (Fig. 6B), the peak moves to Class II, those with two to four individuals per species. The average for two 4-year samples, with 16,000 individuals in 244 species (Fig. 6C), has its peak in Class III. The total of 8 years, all in trap A, gives 32,853 individuals of 285 species, with the peak in Class IV. Finally the total of all traps over 16 trap-years up to 1950 (Fig. 6E) gives 87,400 individuals in 346 species with the peak in Classes IV–V, which includes a range of abundance from 41 to 364 individuals per species.

In 1951 a similar trap, but with a much more attractive mercury-vapour light source, rich in ultra-violet, was used for 9 months and produced 26,300 moths of 265 species, by far the largest sample within a similar period. The number of species in successive × 3 classes was as follows: 31, 45, 55, **65**, 43, 17, 4, 4, 1, with the peak in Class IV corresponding in position with the 8-year sample in trap A with 33,000 insects.

It will therefore be seen that not only does the peak of distribution on this geometric scale move steadily, with increasing sample size, towards greater abundance of species, but there is no evidence that the numbers of species in Groups II, III and above are limited to the 10 per cent above Class I, as required by the logarithmic series. This evidence therefore supports the idea of a log-normal distribution, although the log series may still hold for small samples and for other population data. This will be discussed later.

Dr C. I. Bliss has sent me information of calculations that he has made on my catches of Macro-Lepidoptera at Rothamsted in the years 1933 to 1936. He comes to the same conclusion, that the log-normal gives a better fit than the log series: his results will be published in Bliss (in the press).

PRESTON'S "VEIL LINE" THEORY

When Preston (1948) suggested the log-normal distribution for species with different numbers of individuals in a mixed wild population, he also stated that samples from such a population would produce a log-normal distribution of the same standard deviation, but with a lower median. Thus as the samples became smaller and smaller the curve of distribution would be more truncate at the lower end, as an increasing portion of the theoretical distribution passed below the value of 0 for log individuals per species; that is to say, species which could only have been represented in the sample by a fraction of an individual. This line he called the "veil line" and the species below it are not represented in this particular sample, but might have been in another. All the area of the curve below this point represents a zero term which is really a summation of all the possible fractional occurrences.

To test this against observed distributions, Fig. 7 has been prepared in which the frequency distributions (on a log × 3 scale) for samples of various sizes from the catches of Lepidoptera in trap A at Rothamsted are placed on one another with the zero lines shifted to correspond with the log size of the sample. The smallest sample (1/8 year) had 490 individuals and the largest 32,853 (see Table 9). Over the whole is placed a theoretical log-normal distribution with a standard deviation of 0.95.

It will be seen that the curves fit closely to one another, particularly at the upper end, and that the positions of the n_1 values just above the veil, follow closely to Preston's suggestion for truncate distributions. There appears to be also a steady increase in the height of the peak, showing particularly in the 8-year sample.

More theoretical study and more carefully selected field data are required

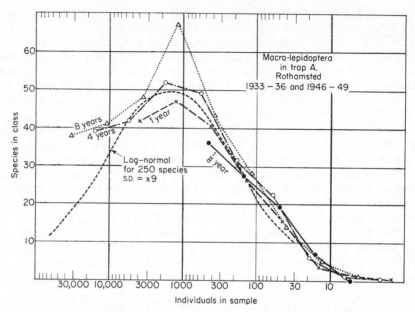

FIG. 7. Frequency distributions of species with different numbers of individuals (× 3 log classes) of Macro-Lepidoptera caught in a light trap at Rothamsted in different periods from 1/8 year to 8 years. The resulting distributions are arranged so that the position of Class 1 varies according to the size of samples on a horizontal logarithmic scale reading from right to left. The curves are superimposed on a log-normal distribution calculated from 250 species, and illustrate Preston's theory of sampling from a population which has a log-normal distribution.

to follow up these points. A possible source of error in the present data is that the increased size of sample was obtained by increasing the duration of the trapping period. During the longer periods there would undoubtedly be an increase in both population and environmental diversity. The use of quadrats in the study of plant populations introduces a similar error, as a larger number of quadrats covers a larger area with a greater likelihood of increased diversity of environment. The best test would be a number of samples taken simultaneously from a large uniform population, which could be combined into groups of varying size. It seems possible that plankton samples might prove suitable for a more critical test, if sufficiently accurate counts could be made, and if a sufficiently high proportion of the individuals could be sorted into species, even if the exact name could not be given to all these.

LEPIDOPTERA IN A LIGHT TRAP IN MAINE, U.S.A.

As an example of a large collection of insects caught in a light trap in another part of the world, we can take data published by Dirks (1937) on 56,131 moths belonging to 349 species captured during 4 years' trapping in Maine, U.S.A.

TABLE 10. *Distribution of species of Lepidoptera with different numbers of individuals, captured in light traps in Maine, U.S.A., during a period of 4 years*

Individ. per sp.	Species	Individ. per sp.	Species	Individ. per sp.	Species
1	38	8	13	15	2
2	36	9	7	16	2
3	19	10	7	17	2
4	10	11	4	18	5
5	14	12	8	19	2
6	13	13	6	20	1
7	11	14	6		

Distribution in × 3 geometric classes

I	II	III	IV	V	VI	VII	VIII	IX	X
38	65	83	54	48	36	19	2	3	1

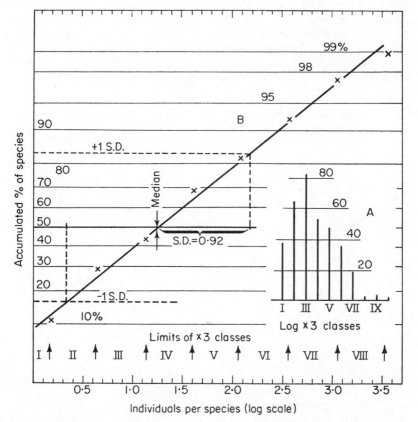

Individuals per species (log scale)

FIG. 8. Frequency distribution of Lepidoptera captured in a light trap in Maine, U.S.A. A = frequency distribution, × 3 log classes of abundance; B = accumulated percentage of species on a probability scale against number of individuals per species on a log scale.

The number of species with up to twenty individuals, and all the species in $\times$ 3 classes are given in Table 10. The highest number of individuals for a single species was 11,424 or 20 per cent of the total captures.

Figure 8 shows diagrammatically the frequency pattern of the $\times$ 3 classes, and also the accumulated percentage totals plotted against a probability scale. The close fit of these points to a straight line suggests a log-normal distribution with the median, on the log scale, at 1.25 and a standard deviation of 0.92. On the arithmetic scale this gives the median at 17.9 individuals per species and a S.D. of 17.8 $\times / \div$ 8.3.

SINGLE-NIGHT CATCHES OF LEPIDOPTERA IN LIGHT TRAPS

To get small samples for this study of distribution we can consider catches in a light trap on a single night. These short-time catches also have the interest of being taken from a cross-section of the population at a more definite moment. They are, however, subject to a large individual sampling error, apart from that due to the relatively small number of species usually captured.

Table 11 and Fig. 9 give particulars of two large catches of Macro-Lepidoptera caught on single nights in light trap "B" at Rothamsted Experimental Station. A photograph of one complete catch is shown in Plate I.

TABLE 11. *Lepidoptera captured in 2 single nights in a light trap at Rothamsted Experimental Station, showing the frequency distribution of species with different numbers of individuals in arithmetic and geometric classes (see also Plate I)*

Individ. per sp.	23 July 1946		9 August 1946	
	obs.	log series	obs.	log series
1	18	15.1	36	29.8
2	8	7.0	8	13.1
3	4	4.4	8	7.6
4	3	3.3	6	5.1
5	—		4	
6	2		2	
7	—		1	
8	1		1	
9	1		—	
10	1		—	
And at	15, 21, 24, and 62		14, 15, 21, and 45	
Total species	42		70	
Total individuals	219		242	

In $\times$ 3 geometric classes

I (1)	18	15.1	36	29.8
II (2–4)	15	14.7	22	25.8
III (5–13)	5⎫		8⎫	
IV (14–40)	3⎬9	⎬12.2	3⎬12	⎬14.4
V (40–121)	1⎭		1⎭	

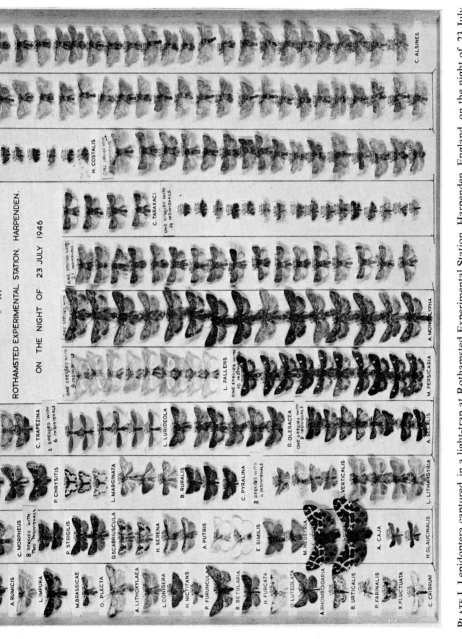

PLATE I. Lepidoptera captured in a light-trap at Rothamsted Experimental Station, Harpenden, England, on the night of 23 July 1946; arranged with the species in sequence of the number of individuals (see Table 11 and Fig. 9) and indicating the typical hollow curve of frequency distribution, with a steady fall in number of species as the number of individuals per species increases. The total catch was 219 individuals representing forty-two species.

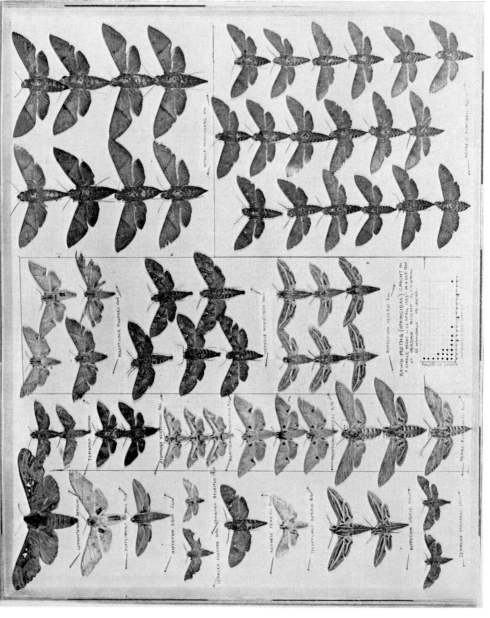

PLATE II. Hawk moths (Sphingidae) captured in a light trap in a single night, 22 April 1953, near Ibadan, Nigeria; arranged in

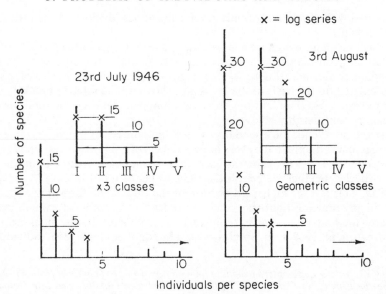

FIG. 9. Frequency distribution of species of Lepidoptera with different numbers of individuals, from light trap catches at Rothamsted, Harpenden, England, on the nights of 23 July (see also Plate I) and 9 August 1946; with arithmetic and log × 3 classes.

On the night of 23–24 July 1946 there were 219 moths belonging to forty-two species. A log-series calculation to fit this number of species and individuals gives an n_1 of 15, somewhat below the observed 18. The fit of $n_2 - n_4$ is close at 14.7 to the observed 15. On the night of 9–10 August, 242 individuals of seventy species were captured. The observed n_1 was 36 as compared with 29.8 for the log series, again slightly too low an estimate. The calculated n_2 is high, partly offsetting the shortage in n_1; the calculated n_3 and n_4 are both within a single unit of the observed. The high error of the single night's catch prevents any more definite conclusions.

LEPIDOPTERA IN SUCTION TRAPS IN ENGLAND

As an example of samples of populations of Macro-Lepidoptera obtained in the field, but not by the use of light traps, we can take the data given by Taylor and Carter (1961) of captures at Cardington, Bedfordshire, England, between the middle of July and the end of October in 3 years, 1954, 1955 and 1959, in four suction traps at heights above the ground of 9, 21 and 56 ft; at the latter height there were two traps. Some of their results are shown in Table 12 and in Fig. 10. The captures are small both in individuals and species, averaging only just over one individual per trap per night. During the 1959 trapping period a catch of 855 individuals of fifty species (diversity 11.2) was obtained in a light trap at Cardington, at about 4 ft above the ground, as compared with 405 moths of only seventeen species (diversity 3.6) in the four suction traps.

It will be seen that—unlike most other population samples of Lepidoptera —there are considerably too many rare species and a few species much too abundant to fit into a log series or a log-normal distribution. The values of n_1 calculated on the basis of the log series, shown in the table in brackets, are in three cases less than half the observed numbers, and even the total for the 3 years is considerably below.

At the other end, the abundant species, the number of individuals in each total catch was dominated by very large numbers of a single species, *Amphipyra tragopogonis* (L.). This came in each year in numbers greater than all the other species together: 72 per cent in 1954, 63 per cent in 1955 and 88 per cent in 1959. This is a far higher proportion of the total catch for a single species than in any previous light-trap catch, or, indeed, any sample of animals made by any method in a similar temperate environment. In the light trap at Cardington in 1959 four species were abundant and the 855 individuals

TABLE 12. *Macro-Lepidoptera caught in four suction traps, at heights of from 4 to 56 ft above the ground, at Cardington, Bedfordshire, between mid-July and late October, in 3 years*

Individ. per sp.	1954	1955	1954–55	1959	All 3 years
1	9 (5.2)	11 (4.8)	17 (7.8)	7 (3.5)	10 (7.0)
2	3	4	7	4	10
3	1	—	2	1	2
4	2	1	2	1	5
5	3	1	1	1	1
6	—	—	—	—	—
7	—	1	—	—	—
8	—	1	—	—	—
9	—	—	1	1	2
10	—	—	1	—	—
11	1	—	—	—	—
12	—	—	—	—	—
13	1	—	—	—	—
and at	23, 225	30, 88, and 269	18, 21, 30, 111, 484	14 and 355	18, 24, 30, 35, 114 and 849
Total individuals	313	430	743	405	1148
Total species	22	22	36	17	36
× 3 Geometric classes					
I (1)	9	11	17	7	7
II (2–4)	6	5	11	6	17
III (5–13)	5	3	3	2	3
IV (14–40)	1	1	3	1	4
V (41–121)	—	1	1	—	1
VI (122–364)	1	1	—	1	—
VII (365–1093)	—	—	1	—	1

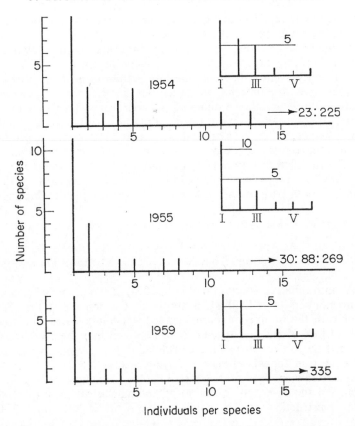

Fig. 10. Lepidoptera caught in a suction trap at Cardington, Bedfordshire, England, in 1954, 1955, and 1959.

included 147 (17 per cent) of *Luperina testacea* (Shiff.), 134 (15.7 per cent) of *Amathes xanthographa* (Schiff.), 128 (15 per cent) of *Tholera popularis* (Fabr.) and 127 (14.9 per cent) of *Amathes c-nigrum* (L.). *A. tragopogonis*, which completely dominated the suction-trap catches, was represented by only a single individual! In the light trap the four species made up 63 per cent of the total, which is a much more normal figure.

It is known that light traps are to a certain extent selective (some species being attracted in a higher proportion than others) but, while this affects the relative proportion of particular species which enter the trap, it should not affect the pattern of the frequency distribution unless there were an association, positive or negative, between abundance and attraction to light. It is almost impossible to imagine that this is so, as a species may be abundant in one locality, or at one season, and rare at another. Any general association of attraction to light with actual abundance would necessitate that individuals would alter their reaction to light according to the number of other individuals of the same species in the district.

Since in the case under discussion the pattern of relative abundance is different in the two traps, and that of the suction trap is very different from samples taken by many methods, the interpretation must be, not that one trap is more free from bias than the other, but that they are sampling different populations, each within the limits of the area from which they draw. The light trap, at about 4 ft from the ground, draws its sample from an unknown but relatively large area, both around the trap and above. The selection made by the suction trap is much more circumscribed and Taylor's results suggest that, while the suction trap is less differentially selective among the species, its very localized range, particularly in height, may be the cause of the great difference in both pattern and composition of the sample obtained.

HAWK-MOTHS (SPHINGIDAE) IN WEST AFRICA

On the night of 22 April 1953 I made a collection of insects, by means of a mercury-vapour light trap, from 8 to 11 p.m. at Moor Plantation, near Ibadan, Nigeria. Sixty-five hawk-moths were captured, which were later classified into nineteen species (Williams, 1954). The number of species with different numbers of individuals is shown in the first column of Table 13 together with the numbers expected on a logarithmic series with the same number of individuals and species. It will be seen that there is a very close fit between the observed and the calculated values. A photograph of this night's collection appears in Plate II (facing p. 33).

In the following year at Kumasi, Ghana, in company with Mr John Bowden, I made a second collection of Sphingidae with a trap of slightly different construction, for the whole of the night of 3–4 June 1954. This

TABLE 13. *Hawk-Moths (Sphingidae) captured in single nights at Ibadan, Nigeria (22 April 1953), and at Kumasi, Ghana (3 June 1954), with estimates made on the basis of the log-series distribution*

Individ. per sp.	Ibadan, Nigeria 1953 observed	calculated	Kumasi,Ghana 1954 observed	calculated
1	7	8.1	7	7.8
2	4	3.6	3	3.6
3	3	2.1	2	2.2
4	1	1.4	1	1.5
5	1	1.0	2	0.95
6	1	0.7	1	0.84
7	—		2	
8	1		1	
12	—		1	
17	—		1	
18	1		—	
Total individuals	65		92	
Total species	19		21	

collection included ninety-two individuals of twenty-one species, with the distribution shown in the third column of Table 13, again with the numbers expected on the basis of a logarithmic series. Once again the fit is very close. The two sets of observations are shown diagrammatically in Fig. 11.

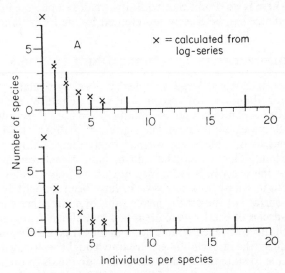

FIG. 11. Frequency distribution of species of Sphingidae (Hawk-Moths) captured in light traps in West Africa. A, captured near Ibadan, Nigeria, on the night of 22 April 1953; B, captured at Kumasi, Ghana, on the night of 3 to 4 June 1954. The crosses show the numbers expected on the basis of a logarithmic series.

TIPULIDAE (DIPTERA) CAPTURED IN A LIGHT TRAP AT
HAILEYBURY COLLEGE, HERTFORDSHIRE, ENGLAND

In the report of the Haileybury School Natural History Society for 1951 there is a list of twenty-eight species of Tipulidae captured in a light trap. The numbers of species with different numbers of individuals is given in Table 14.

TABLE 14. *Tipulidae (Diptera) captured in a light trap at Hailey-bury College, Hertfordshire, England, in 1951*

Individuals	Species	Individuals	Species
1	9	7	3
2	3	11	1
3	2	23	1
4	4	25	1
5	2	59	1
6	1		
Total individuals = 192; total species = 28			

The arithmetic mean individuals per species was 6.85. On the assumption of a logarithmic distribution, the estimated number of species with one individual is 8.6, as compared with the observed value of 9. The estimated number with one to three individuals is also 8.6 as compared with the observed 9. There is no doubt that with this small sample, the distribution of relative abundance can be closely represented by the logarithmic series.

COLEOPTERA IN SEA-SHORE DRIFT IN FINLAND

Apart from trap catches, large samples of insect populations which are randomized for abundance of species, and in which all or nearly all of the species have been identified and the numbers separately recorded, are difficult to find. Some examples have been given by Palmén (1944) of samples from great masses of Coleoptera washed in by the sea along the southern shores of Finland. They cannot be said to have come from one particular association, as their source is unknown, nor do we know if they were originally washed out to sea by flooded rivers, or have been brought down from the air on to the surface of the sea by heavy rain, or some other cause.

Over 970 species of beetles were identified by Palmén in nine such aggregations. For each species in each sample he gives the actual number of individuals if it is below fifty; the larger numbers are only roughly estimated. As a result it is not possible to state the exact number of individuals in the sample. Two of these samples are discussed in Williams (1953a) and details of one, which had a total of 466 species, are given in Table 15 and Fig. 12.

TABLE 15. *Coleoptera taken from great masses of insects washed up on the sea-shore in S. Finland by Palmén*

Individ. per sp.	No. sp.	Individ. per sp.	No. sp.	Individ. per sp.	No. sp.	Individ. per sp.	No. sp.
1	111 (111)	11	3	21	2	34	1
2	87	12	7	22	1	36	2 (37)
3	59	13	4 (99)	23	1	49	1
4	35 (181)	14	5	24	1	50	1
5	29	15	5	25	1	50–100	11
6	20	16	4	29	2	100–200	9
7	6	17	5	30	1	200–500	4
8	10	18	1	31	2	500–1000	1
9	17	19	1	32	1	over 1000	1
10	3	20	0	33	1	"infinite"	10

It will be seen that in relation to the number of species with one individual, n_1, there are far too many species with two or more individuals than could be accounted for by a logarithmic series. The probability scale diagram in the figure also rules out the log-normal distribution, as the line is very definitely curved. Nor is it near a Poisson distribution. Some special conditions in the unknown causes of selection must contribute to this somewhat unusual distribution.

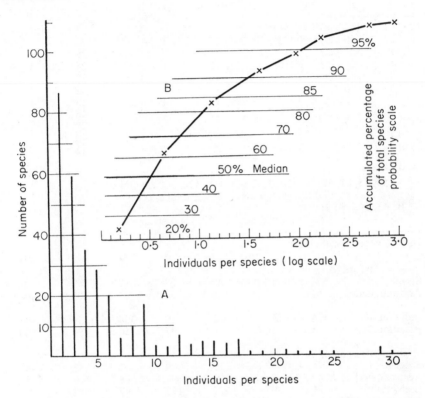

FIG. 12. Coleoptera collected from sea-shore drift in Finland, showing the frequency distribution of species and different numbers of individuals. A shows an arithmetic scale of abundance; B, using a log scale of abundance and a probability scale of accumulated percentage species, shows a strongly curved relation, excluding the possibility of a log-normal distribution.

COLEOPTERA IN RIVER-FLOOD REFUSE IN ENGLAND

Easton (1947) gave particulars of two collections of beetles obtained from flood refuse on the banks of rivers. The first (A) was from the River Thames at Oxford on 1 December 1946; the second (B) from the River Mole near Mickleham, Surrey, on 25 November of the same year.

The numbers of species with different numbers of individuals are shown in Table 16 and in Fig. 13. There is in each case the typical hollow curve, but calculations on the basis of the log series show in both a large excess of the observed species with one individual over that calculated. The figures are:

TABLE 16. *Frequency distribution of beetles from samples of river-flood refuse.* A, *on the River Thames at Oxford.* B, *on the River Mole near Mickleham, Surrey*

Individ. per sp.	Species A	B	Individ. per sp.	Species A	B	Individ. per sp.	Species A	B
1	114 (68)*	71 (46)*	12	4	4	24	1	—
2	40 (29)	20 (22)	13	3	—	25	1	—
3	24 (22)	11 (14)	14	—	1	26	1	—
4	16 (16)	6 (10)	15	1	2	28	1	—
5	13	7	16	3	1	30	1	—
6	6	7	17	1	—	32	1	—
7	3	3	19	4	1	33	1	—
8	8	2	20	1	2	36	—	1
9	2	2	21	1	—	38	1	—
10	2	2	22	—	1	40	1	—
11	1	3	23	2	—	42	2	—

And also at

A: 53, 55, 57, 69, 76, 124, 125, 144, 164, 174, 200 (2) and 406
    Total individuals = 3102; total species = 273
B: 43, 54, 134 and 158
    Total individuals = 957; total species = 150

In × 3 geometrical classes

	I	II	III	IV	V	VI	VII
A = Oxford							
Species	114 (68)*	80 (66)*	42	22	7	7	1
Accumulated %	41.8	71.1	86.4	94.5	97.1	99.6	100
Individuals	114	216	322	519	394	1131	406
Accumulated %	3.7	10.6	21.0	37.7	50.5	86.9	100
B = Mickleham, Surrey							
Species	71 (46)*	37 (46)*	29	9	2	2	—
Accumulated %	47.3	72.0	91.3	97.3	98.7	100	
Individuals	71	97	222	177	97	292	
Accumulated %	7.4	17.6	40.8	59.2	83.3	100	

* Calculated on log series.

Oxford, observed 114, calculated sixty-eight; Surrey, observed seventy-one, calculated forty-six. In the Oxford sample this excess persists in n_2 and n_3: in the Surrey sample there are fewer observed than calculated in all the n_2 to n_4 groups, but insufficient to make up the excess in n_1. There must, therefore, be a deficit in the more abundant species to make up the balance.

This difference may be due to a different mathematical pattern, but it may also be due to a biased sampling error. Mr Easton tells me that not all the individuals in the sample were identified, and that those not included in the count were more likely to be the commoner species. In some of these, particularly when very small, a large number were not identified with certainty. This source of error exists in nearly all collections not specifically made for statistical analysis.

The table also shows the data sorted into × 3 classes, and the accumulated percentage totals for both species and individuals. It appears that in the Oxford collection the rarer 50 per cent of the species account for only about 6 per cent of the individuals, while the single most abundant species, with 406

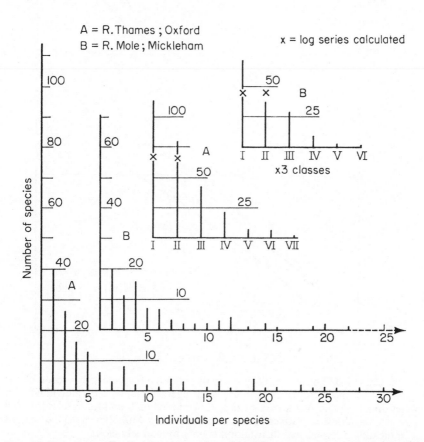

FIG. 13. Frequency distribution of species with different numbers of individuals in two samples of Coleoptera from river-flood refuse in England. A = River Thames; B = River Mole. Each distribution is shown on an arithmetic and a × 3 log scale of abundance.

individuals, accounts for 13.1 per cent, this latter probably being an underestimation. In the Surrey sample the 50 per cent rarer species account for 8 per cent of the individuals, and the most abundant species (158 individuals) for 16.5 per cent.

The transfer of the accumulated percentages to a probability graph is shown on Fig. 14 and indicates only an approximate fit to straight lines of a log-normal distribution.

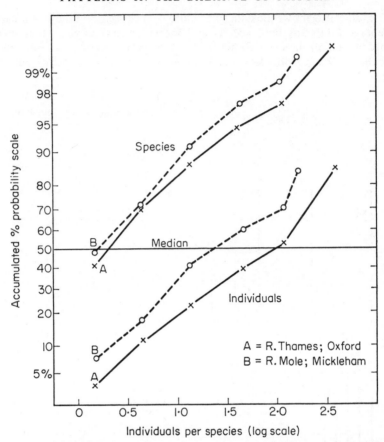

Fig. 14. Data as in Fig. 13 for Coleoptera in river-flood refuse, plotted as accumulated percentage of species on a probability scale against numbers of individuals per species on a log scale. Values are given both for individuals and for species for each of the two localities. A = River Thames, and B = River Mole. The fit to the straight line indicating a log-normal distribution is only moderately close.

CARABIDAE (COLEOPTERA) IN GROUND TRAPS IN HOLLAND

A collection of insects made by an entirely different method is recorded by Den Boer (1958), who captured a large number of Carabid beetles in 100 "bucket traps" let into the ground on sand-dunes in Holland during the 12 months 1 March 1953 to the end of February 1954. His results, including some additional details obtained direct from the author, are summarized in Table 17 and Fig. 15.

There is the usual hollow-curve type of distribution. The number of species observed with only one individual (13) is above that estimated from the log series (9.5). For the species with two to four individuals the fit is closer,

TABLE 17. *Carabid beetles captured in 100 traps in sand-dunes in Holland from 1 March 1953 to 1 March 1954 by P. J. Den Boer, with log-series estimates for comparison*

Individ. per sp.	Species	Individ. per sp.	Species	Individ. per sp.	Species
1	13 (9.5)	6	2	15	1
2	2 (4.7)	7	1	21	2
3	2 (3.2)	8	4	25	1
4	5 (2.4)	12	1	30	1
5	3	14	1	31	1

And also single species at:

35, 36, 44, 49, 60, 65, 72, 76, 87, 104, 152, 180, 287, 320, 351, 465, 496, 510, 831, 881, 983 and 1721

Total individuals = 8113; total species = 63

In × 3 classes

I (1)	13 (9.5)	V (41–121)	8
II (2–4)	9 (10.3)	VI (122–364)	5
III (5–13)	11	VII (365–1093)	6
IV (14–40)	10	VIII (1094–3280)	1

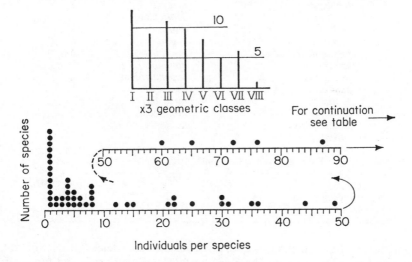

FIG. 15. Carabidae (Coleoptera) captured in sand-dunes in Holland, showing the frequency distribution of species with different numbers of individuals, on an arithmetic scale and on a × 3 geometric scale.

9 observed and 10.3 calculated. The most abundant species, *Calathus erratus* Sahlb., accounts for 21.2 per cent of the total individuals, and the four commonest species comprise over 55 per cent of the total population. At the other end of the scale the thirty-two least common species (50 per cent of the

total species) account for only 1.3 per cent of the total individuals. There are too many abundant species to make a good fit to either the log-normal or the log-series distributions.

DIPTERA IN COW-DUNG IN S.E. ENGLAND

An interesting set of data in which 3000 individuals of a single genus of Diptera, Limosina (Borboridae), were collected was given by Laurence (1955). The flies were collected on cow-pats at Rothamsted Experimental Station (about 25 miles north of London) each month from October 1950 to September 1951. The total numbers of each species found is given in Table 18, which also includes the number of species in each of the × 3 classes for individuals per species.

There are too many species with one individual each to be a good fit to a

TABLE 18. *Borboridae (Diptera) of the genus* Limosina *captured in cow-dung from October 1950 to September 1951 at Rothamsted Experimental Station by D. B. Laurence*

Species of the genus *Limosina*, Diptera, Borboridae

scutellaris	1015	*vitripennis*	4
lugubris	902	*moesta*	3
crassimana	692	*denticulata*	3
collinsi	167	*atoma*	3
sylvatica	114	*vagans*	2
pseudo-leucoptera	63	*mirabilis*	2
spinipennis	18	*humida*	1
ferrugata		*appendiculata*	1
pusilla	8	*flaviceps*	1
fontinalis	7	*hirticula*	1
aterrima	5	*minuscula*	1
quadrispina	4	sp. *indet.*	1
rufilabris	4		

Total individuals = 3035; total species = 25

× 3 classes	I	II	III	IV	V	VI	VII
No. of species	6	8	4	1	2	1	3
Log series	3.8	4.0					

logarithmic series (which gives 3.8), nor does the geometric classification give a good fit to a log normal. There are, in fact, too many rare and too many common species to fit either of these suggested distributions.

SOIL COLLEMBOLA IN THE ALPS

Agrell (1941) gives particulars of certain samples of Collembola taken from dead pine leaves (A), dead leaves (B), and moss on stones (C) in an Alpine situation. The number of individuals per species in these samples are as follows:

A: 1, 1, 1, 2, 2, 3, 4, 4, 5, 11, 17, 19, 35, 148.
253 individuals of fourteen species
B: 1, 1, 1, 1, 2, 3, 3, 5, 6, 7, 9, 10, 11, 12, 15, 33, 34, 68, 81.
303 individuals of nineteen species
C: 3, 3, 21, 31, 37, 120.
215 individuals of six species.

On the assumption of the logarithmic series the calculated numbers of species with one individual, as compared with the observed, are:

A: 3.0 to 3, B: 4.2 to 4, C: 1.1 to 0,

showing that this series gives quite close estimates of the value of n_1.

SPIDERS TAKEN FROM THE AIR IN A NET IN ENGLAND

Freeman (1946, p. 70) gave a list of forty-three individuals of twenty species of spiders caught in nets attached to wireless masts up to a height of 300 ft at Tetney, Lincolnshire, England, in 1934 and 1935.

The frequency distribution of species with different numbers of individuals was:

Individuals per sp. 1, 2, 3, 4, 5, 6.
Species   11, 2, 3, 2, 1, 1.

This is of the familiar hollow-curve distribution and a calculation of n_1 on the assumption of the logarithmic series gives 10.75, as compared with the observed 11.

ABUNDANCE OF BRITISH NESTING BIRDS

To obtain figures for a much larger population I asked James Fisher if it was possible to place the known nesting land-birds of England and Wales into classes of abundance in powers of 10. That is to say to put each known species according to the average number of birds present during the breeding season in classes of 1–10, 10–100, 100–1000, etc. (see Fisher, 1952, and Williams, 1953a). The number of species considered was 142, with a total population of about 63 million individuals; the number of species in each class is shown in Table 19. The placing was partly on census work, particularly for the rarer species, and partly on the estimates made from the known area of the distribution on habitat, and the approximate number of individuals per acre.

The results are shown graphically in Fig. 16, which includes a plotting of the class sizes against a probability scale. Both figures support the log-normal distribution, and the probability scale suggests that it has a mean log of 3.75 and a standard deviation of ± 1.55. This implies a geometric mean of about 5600 individuals per species, and as this is also the median, about half the species will be more abundant and half less abundant than this. The arithmetic mean of about 440,000 individuals per species is far above the median, due to the geometric form of the variation. The standard deviation of the

TABLE 19. *Frequency distribution of 142 species of land-nesting British birds in × 10 classes of abundance, from estimates made by J. Fisher. Also for comparison, calculations made on the basis of a log-normal distribution.*

Individ. per sp.	Numbers of species Fisher's estimate	From log-normal
1–10	7	5.7
10–100	9	12.8
100–1000	22	26.9
1–10,000	42	34.1
10–100,000	32	32.7
100,000–1 million	16	19.9
1–10 million	12	7.3
Over 10 million	2 (just over)	2.2
Over 100 million	0	0.4

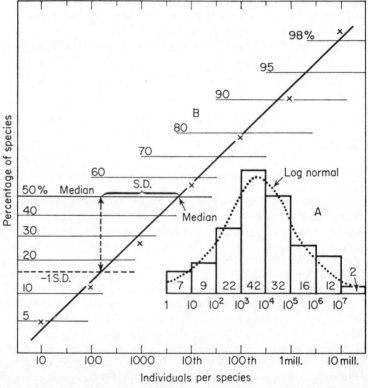

FIG. 16. Frequency distribution of estimated abundance of 142 land-nesting species of birds in England and Wales. A, in × 10 log classes; B, with accumulated percentage of total species plotted against abundance of individuals per species. The latter gives a very close fit to a straight line, indicating a log-normal distribution with a median at about 5600 individuals per species.

log normal also suggests that approximately two-thirds of the species lie between the limits of 3.75 $\pm$ 1.55 on the log scale, or 7000 $\times$/ $\div$ 35 on the arithmetic scale; that is to say, between 160 and 200,000 individuals per species. Approximately one-sixth of the species will be above the upper limit, and one-sixth below the lower limit.

FIELD COLLECTIONS OF SNAKES IN PANAMA

In 1949 Dunn and Allandoerfer published an analysis of the number of individuals per species in several sets of field collections of snakes in Panama, and discussed this in relation to the logarithmic series. In two sets of data the material was from collections made for museums and was possibly biased in favour of the rarer species; the other four collections were more random. In Table 20 the information for these four collections is tabulated in $\times$ 3 geometric classes (with some additional data obtained direct from Mr Dunn), and the table and Fig. 17 also show the average number of species per class in the four collections together. It will be seen that in three out of the four collections the observed number of species with one individual is below that estimated from the logarithmic series, and in collection D very considerably below.

TABLE 20. *Frequency distribution of species of snakes with different numbers of individuals from four collections made in Panama by Dunn and Allandoerfer*

Class and individ. per sp.		A Cocle	B Sabanas	C Darien	D Chagres	Average all four
I	1	6	6	5	3	5.0
II	2–4	5	8	7	9	7.25
III	5–13	3	14	4	16	9.25
IV	14–40	9	9	13	8	9.75
V	41–121	4	4	10	10	7.0
VI	122–364	3	9	3	7	5.5
VII	365–1093	—	2	2	—	1.0
Total species		30	52	44	53	
Total individuals		1232	3914	3041	2500	
n_1 calculated for log series		5.5	8.5	7.3	9.5	7.7

To test the alternative fit to a log-normal distribution, suggested by the symmetrical form of Fig. 17A, the accumulated totals were plotted against a probability scale in Fig. 17B. The straight-line relation, except at the upper end, indicates a very close fit to a log-normal with the median, on the log scale, at approximately 1.13 (13.5 individuals per species), and a standard deviation of 0.85 (which is equivalent to a multiplication or division by 7).

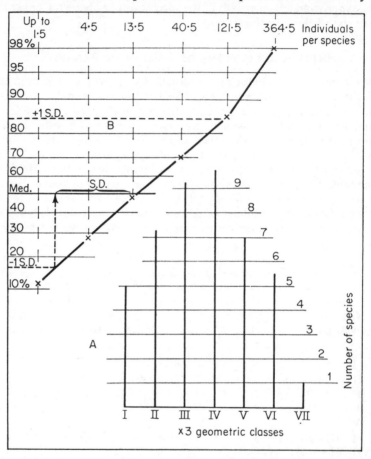

FIG. 17. Frequency distribution of species of snakes with different numbers of individuals, from an average of four field collections in Panama. A, with × 3 geometric classes of abundance; B shows the same data plotted against a probability scale of accumulated percentage species. The straight line indicates a close fit to a log-normal distribution with a median at 13.5 individuals per species. The upper point at 98 per cent is subject to considerable error.

NESTING BIRDS IN QUAKER RUN VALLEY, NEW YORK STATE

In 1936 Saunders gave an account of the nesting-bird population in the Quaker Run Valley, Allegany State Park, New York, which has an area of about 26.5 square miles (38 km² or 17,000 acres). The number of species of birds with different numbers of nesting pairs is given in Table 21. The total number

TABLE 21. *Frequency distribution of species of birds with different numbers of nesting pairs in Quaker Run Valley, New York State, U.S.A.*

Pairs per sp.	No. of sp.	Species with the following number pairs; one species only unless indicated			
1	2	15	35	188	506
2	1	17	43 (2)	220	675
3	4	22	46 (2)	270	723 (2)
4	4	23	50	280	868
5	1	24	56	282	1196
6	2	26	57	288	1656
7	1	28 (2)	60	310	1670
8	2	30	79	311	
10	5	32 (2)	88	324	
12	1	33	90	389	
14	1	34	91	477	

Total species = 79; total pairs = 14,350

In geometric × 3 classes

Class (pairs)		Sp. in class	As accum.	Total pairs in class	As accum.
		No.	%	No.	%
I	(1)	2	2.5	2	0.014
II	(2–4)	9	13.9	30	0.22
III	(5–13)	12	29.1	102	0.93
IV	(14–40)	15	48.1	393	3.7
V	(41–121)	14	65.8	969	10.4
VI	(122–364)	16	86.1	3604	35.5
VII	(365–1093)	8	96.2	4728	68.5
VIII	(1094–3280)	3	100	4522	100

of species was seventy-nine and the number of pairs 14,350; which gives a density of just under one nesting pair per acre. The average (arithmetic mean) number of pairs per species is 181.6. When arranged in × 3 geometric classes the distribution (Fig. 18) resembles a log normal, and a test against a probability scale, shown in the same figure, shows a close fit to a log normal with a median, on the log scale, of pairs per species of 1.65 + 0.8, or, on the arithmetic scale, 44.7 ×/ ÷ 6.3. The median is, as expected from the form of the distribution, considerably below the arithmetic mean. The relative proportions of the population made up by the rare and by the most abundant species is discussed on p. 110.

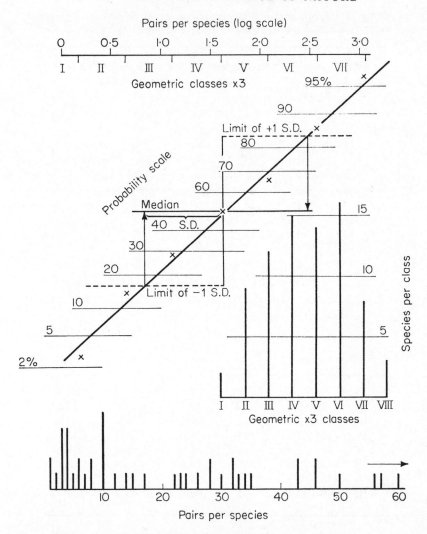

Fɪɢ. 18. Birds in Quaker Run Valley, New York State. Frequency distribution (in pairs of individuals per species) on arithmetic and × 3 log classes, and also plotted against a probability scale of percentage of species. The latter indicates a moderately close fit to a log-normal distribution with a median at approximately 40 (log 1.6) pairs per species.

Tʜᴇ Iʟʟɪɴᴏɪs Cʜʀɪsᴛᴍᴀs Bɪʀᴅ Cᴇɴsᴜs

For many years a bird census in mid-winter has been taken in the State of Illinois, U.S.A., by a number of observers at a number of localities. I have taken the data for the 4 years 1954–57 (see References: Illinois Christmas

Bird Census) to study the frequency distribution of species with different numbers of individuals, and also some other related problems (see p. 63 and p. 282).

A total of 1,145,000 birds were counted belonging to 146 species, of which the most abundant was the Mallard, with over 800,000 individuals, and, at the other end, fifteen species seen each only once. The total number for each species in the 4 years is given in Appendix C, and on Fig. 19 is the frequency

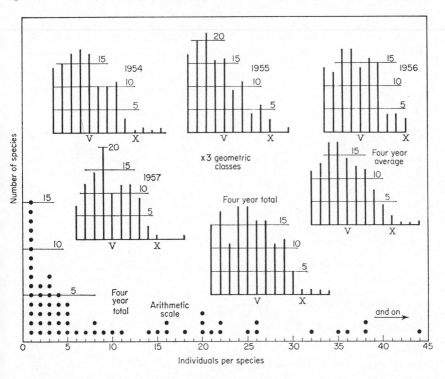

FIG. 19. Frequency distribution of species with different numbers of individuals for American birds from data from the "Illinois Christmas Bird Census", 1954–7. Part of the 4 years' total (up to forty-five individuals per species) is shown on an arithmetic scale. The results for single years, for the 4 years' average and the 4 years' total are shown with × 3 geometric classes of abundance.

distribution of the sixty-three species which had up to forty-five individuals each. Owing to the enormous variation of the Mallard from year to year—from only 4450 in 1956 to 657,637 in 1957—it has been omitted from some of the calculations. Without the Mallard, the total number of birds, 340,196, in 145 species, gives an arithmetic mean of 2346 birds per species, but the median is far below this at only 68. This is, of course, typical of the geometric or logarithmic scale of variation.

On the basis of the log series (Table 22), and again omitting the Mallard, the total 4 years' count gives a diversity of 14.5 and an estimated n_1 of the same number. It is interesting to note that the observed n_1 was 15. For the different years the diversity is very uniform, ranging only from 12.3 to 13.8, indicating considerable uniformity in population structure from year to year. The form of the distribution, when brought to a logarithmic scale, does not, however, support the logarithmic series.

TABLE 22. *Frequency distribution (in* × *3 classes) of species of birds with different numbers of individuals in the Illinois Christmas Bird Censuses from 1954 to 1957*

Class		1954	1955	1956	1957	4 Years Average	Total
I	1–	14	12	14	7	11.75	15
II	2–4	15	15	13	12	13.75	18
III	5–13	17	22	18	15	18.00	11
IV	14–40	18	16	18	20	18.00	19
V	41–121	17	16	14	12	14.75	19
VI	122–364	10	9	16	14	12.25	16
VII	365–1093	10	11	15	14	12.50	18
VIII	1094–3280	11	4	4	11	7.50	10
IX	3281–9841	3	6	4	3	4.25	12
X	9842–29,524	—	3	3	1	1.75	5
XI	29,525–88,573	1	—	—	—	0.25	1
XII	88,574–265,720	—	1	—	—	0.25	1
XIII	265,721–797,161	1	—	—	1	0.50	0
XIV	over 797,161	—	—	—	—	—	1
Total species		117	115	119	110	115.50	146
Individuals excluding mallards		93,707	88,534	69,008	82,591		333,840
Mallards only		107,341	35,375	4453	657,637		804,806
Diversity (α) excluding mallards		13.1	13.2	13.8	12.3	13.1	14.5

In Table 22 the number of species in each of the × 3 classes of abundance is shown for each of the 4 years, for their average, and for the total of all years. The same distributions are shown graphically in Fig. 19. There is some irregularity in each of the single years, and in the total. It is, however, of interest to note the expected rise of the peak frequency from Classes III and IV in the single years and in the average, to IV and V in the 4 years' total.

The frequency distribution from the average of the 4 years (see Fig. 19) is very regular, and resembles a truncate log normal as suggested by Preston. Tests for this using the accumulated percentage and a probability scale are shown in Table 22 and in Fig. 20. The first, A, is calculated by neglecting any possible zero term. This gives an approximately straight-line relation, slightly curved at the lower end and irregular at the extreme upper end, where, however, the error is very high.

It will be seen from Table 21 that the total number of species in the 4 years was 146, and the average of the single years 115.5. The difference of thirty would, however, include some species which were not present in the population in every year. If we assume that half of these were present in each year, but were not observed, we get an estimate of 15 for the zero term in the log normal. On this basis the accumulated percentages in Table 23 and Fig.

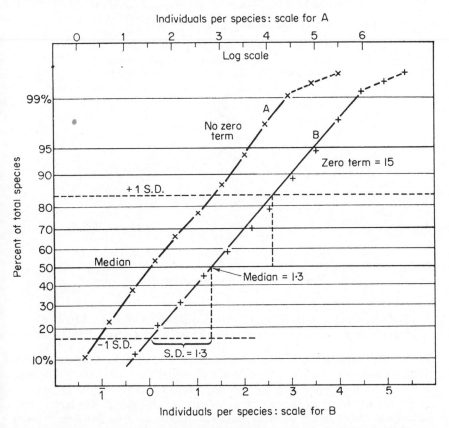

FIG. 20. Frequency distribution of abundance of birds from the "Illinois Christmas Bird Census" (see Fig. 19). A, shown as accumulated percentage of species on a probability scale plotted against increasing numbers of individuals per species, and also, B, as accumulated percentage of individuals. In both cases the relations are close to a straight line, except for the higher values, which are subject to an increasing error. This indicates a close fit to the log-normal distribution.

20B show, except for the upper limits, an almost perfect fit to a straight-line relation. This implies a log-normal distribution which has its median at 1.30 on the log scale, or twenty individuals per species, and a standard deviation of $\pm$ 1.3 on the log scale, or $\times/ \div 20$ on the arithmetic scale.

There is, therefore, from this extensive series of bird surveys, good evidence of a log-normal distribution in the population, with a small zero term including those species which, though present in the population, were not observed in every year.

TABLE 23. *Illinois Christmas Bird Census, 1954–7. Accumulated percentage of species with different numbers of individuals for the average of 4 years, with and without an estimate of the zero term*

× 3 classes	Accumulated %, 4-year average	
	No zero	Zero assumed 15
0	—	11.5
1	10.2	20.5
2	22.1	31.0
3	37.7	44.8
4	53.2	58.6
5	66.0	70.0
6	76.6	79.3
7	87.4	88.9
8	93.9	94.6
9	97.6	97.9
10	99.13	99.24
11	99.35	99.43
12	99.57	99.62
13	100	100

INDIVIDUALS AND SPECIES IN BIRDS RINGED AT MONKS' HOUSE BIRD OBSERVATORY, NORTHUMBERLAND, ENGLAND

One of the difficulties in the analysis of field surveys of large numbers of birds is the varying uncertainty of the identification and the exact numbers of each species. The data published by Ennion (1960) on birds ringed at Monks' House on the coast of Northumberland, in N.E. England, are largely free from these two sources of error, as every individual was handled by very experienced workers. The results, given in detail in Table 24, show that in the years 1951 to 1958, 23,934 birds belonging to 140 species were recorded, with an arithmetic mean of 171 individuals per species.

It is perhaps interesting to note once more, as further evidence for the geometric scale of variation, that 107 species are below the arithmetic mean and only thirty-eight above. The median abundance is twenty-one individuals per species, half the species having fewer and half more than this number, while the geometric mean is at 24.5 individuals per species.

It will be noted that when the species are grouped into × 3 classes of abundance, the number of species in the first six classes are not very different, in the neighbourhood of twenty species in each class. This is different from

TABLE 24. *Numbers of species of birds with different numbers of individuals, ringed at Monks' House, Northumberland, England, between 1951 and 1958 with earlier terms as calculated from the logarithmic series*

1	19 (19)	19.6 (19.6)	15	1	39	1	104	1
2	8	9.89	16	1	40	1 (26)	105	1
3	6	6.51	17	4	42	1	107	1
4	4 (18)	4.88 (21.2)	18	3	44	1	110	1
5	2	3.71	19	2	47	1	115	1
6	7	3.26	21	2	50	1	117	1
7	2	2.78	23	2	53	1	120	1 (19)
8	1	2.43	25	2	57	1	124	1
9	—	2.17	28	2	68	2	151	1
10	4	1.95	30	1	73	1	152	1
11	—	1.77	31	1	89	1	166	1
12	4	1.61	33	1	93	1	175	2
13	1 (21)	1.49 (21.2)	34	2	99	1	193	1

Also single species with: 204, 211, 247, 265, 280, 286, 304, 311, 313, 326, 329, 349, 353, 377, 395, 399, 422, 517, 571, 585, 596, 605, 610, 774, 846, 998, 1253, 1617, 2792 and 3232 individuals.

Total: 23,934 individuals of 140 species.

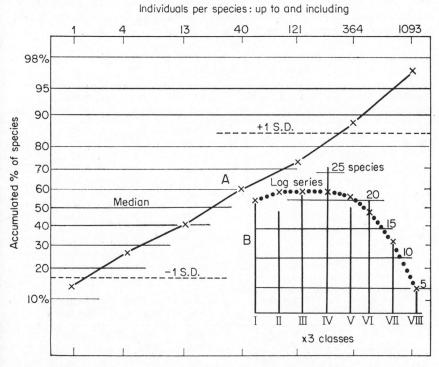

FIG. 21. Birds ringed at Monks' House, Northumberland, England, between 1951 and 1958.

the log-normal distribution and it is seen, in Fig. 21A, that a probability diagram does not give a close fit to a straight line. When a log series is calculated with $x = .99918$ the fit is close, the observed and calculated values for the successive $\times$ 3 classes being:

Class	1	2–4	5–13	14–40	41–121	122–364	365–1093	over 1093
Observed	19	18	21	26	19	20	13	4
Log series	19.7	21.3	21.5	21.2	20.4	18.1	12.7	5.2

These results are shown diagrammatically in Fig. 21B.

This particularly reliable set of field data, with a high average number of units per group, gives a better fit to a log-series than to a log-normal distribution.

DISTRIBUTION OF ABUNDANCE IN FRESH-WATER ALGAE

Margalef (1949a, b) gave details of a number of samples of fresh-water algae from small ponds in N.E. Spain from which Table 25 has been prepared. It will be seen that, with a total of 2348 individuals belonging to 355 species, the typical hollow curve was found (Fig. 22) between the number of species and their abundance. Margalef calculated a log series ($x = 0.95276$ $\alpha = 116.4$) with the results also shown in summary in the table. This series underestimates the number of species with one individual (n_1), is almost correct for n_2, but generally overestimates from n_3 to n_{13}.

TABLE 25. *Frequency distribution of species of fresh-water algae from samples from a number of small ponds in N.E. Spain*

Individ. per sp.	No. of sp.	Individ. per sp.	No. of sp.	Individ. per sp.	No. of sp.	Individ. per sp.	No. of sp.
1	136	10	3	19	1	28	1
2	53	11	7	20	1	29	2
3	24	12	8	21	2	30	3
4	20	13	1	22	3	31	3
5	19	14	3	23	1	32	2
6	11	15	5	24	2	33	1
7	9	16	2	25	1	34	1
8	13	17	2	26	2	over	
9	3	18	1	27	1	35	7

Total species = 355; total individuals = 2348

In geometric $\times$ 3 classes (approximately):

Class range	Species No.	Accum. %	Individ. in class No.	Accum. %	Species on log series
1	136	38.3	136	5.8	110.9
2–4	97	65.6	258	16.8	110.4
5–13	74	86.5	371	41.1	86.2
14–34	41	98.0	970	82.4	38.5
over 35	7	100	413	100	8.9

In Margalef's table (1949b) he does not give separately the numbers of individuals per species above n_{34} but the total number is 7, and the number of individuals in these is, by subtraction, 413. Table 25 also shows the number of species in × 3 geometric groups, as far as possible, and Fig. 21 shows these transferred to a probability diagram. The former shows a distribution falling steadily, which might, however, be a very truncate log normal without any peak. The latter (made without adding any estimated zero term) gives a closer fit to a straight line than might have been expected, with the median at about 2.5 individuals per species.

The opportunity was taken to make the same calculation for the accumu-

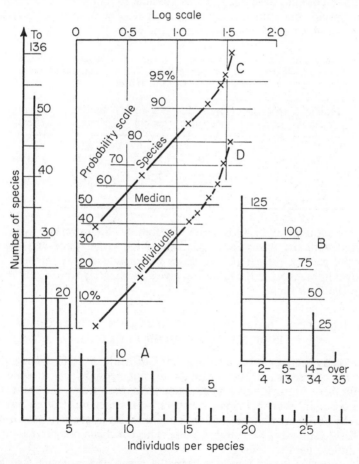

FIG. 22. Frequency of distribution of fresh-water algae in N.E. Spain as measured by numbers of individuals per species. A, on an arithmetic scale; B on × 3 scale; C, abundance plotted against accumulated percentage on a probability scale for species and for individuals. There is an indication of a straight-line relation except in the upper values, where the error is large.

lated percentage of individuals. This shows (Fig. 22D) a line approximately parallel to the first (as expected from the theory of the log normal, see p. 107) with a median at 1.27. Thus half the individuals are in species containing eighteen or fewer individuals.

FUNGUS SPORES IN THE AIR: DISTRIBUTION IN GENERA

In 1949, Hyde and Williams described collections of spores of fungi falling from the air on to petrie dishes on the roof of a building near Cardiff, South Wales. The traps were exposed for 10 minutes every morning for 1 year, and of the 2988 colonies developed, it was possible to identify 2302 to seventeen genera. Specific determinations were not made. The number of genera with different numbers of spores is shown in Table 26.

TABLE 26. *Number of genera of fungi represented by different numbers of spores, in samples falling from the air on to petrie dishes, near Cardiff, Wales*

Individ. per genus	No. of genera	Individ. per genus	No. of genera	Individ. per genus	No. of genera
1	3	12	1	82	1
2	1	20	1	108	1 (3)
4	1 (5)	21	1	380	1
7	2	34	1 (7)	1542	1 (2)
11	1	69	1		

On the assumption of a log-series distribution, the estimated n_1 is 2.53 as compared with the observed 3. Because of the large range and the small number of genera, the geometric classes were made at $\times$ 9 intervals. These are shown in brackets in the table and give a moderately smooth curve with a peak in the 5–40 range.

THE RELATIVE PROPORTIONS OF RARE AND COMMON SPECIES IN A MIXED WILD POPULATION

It has already been mentioned (p. 17) that Garthside in 1928, commenting on his catches of insects in traps in Minnesota, summed up "that, despite of the large number of individuals that characterize some species, the great bulk of species occur in relatively small numbers." This fact must often have been noticed by observant field naturalists, but Garthside appears to have been one of the first to state it clearly, and to support his conclusions by actual numbers from random field samples.

This pattern follows logically from the types of frequency distribution that we have been discussing, but it is of interest to examine the problem on its own, and to see how the patterns may differ in different groups of animals, in different-sized samples, and in samples taken by different methods. In Table 27 some of the relevant data are given for nine samples. In Fig. 23 there is shown,

TABLE 27. *The relative proportion of rare and common species in animal populations, with the proportion of total populations of individuals that they supply. Also the abundance of the single most abundant species as a percentage of the total population. Examples from various animal groups*

Group of animals and type of sample	No. of individ.	No. of sp.	Average individ. sp.	% of individ. for 50% rarer sp.	% of most common sp. for 50% of individ.	Single most abundant species No. of individ.	As % of total individ.
Lepidoptera Light trap							
(A) Rothamsted. 1 year 1935	6815	197	34.6	3.2	3	1799	26.6
(B) Nigeria. 1 night Sphingidae only	65	19	3.4	18.0	16	19	27.7
Suction trap Cardington. 3 months per year							
(C) 1954–1955	743	30	24.7	2.5	<1 sp.	743	65.6
(D) 1959	405	17	23.9	2.5	<1 sp.	405	87.7
Aphidae Suction trap							
(E) Rothamsted	1451	58	25.0	2.4	6	223	22.3
Birds London. 26 counts in field in winter							
(F) Quaker Run Valley, N.Y.	170,462	62	26,632	0.43	<1 sp.	90,000	52.8
Nesting pairs; field estimates (G) England and Wales.	14,350	79	1816	4.9	8	14,310	11.6
Land nesting species, census and estimates (H)	65 mill.	142	458,000	0.3	4	10.5 mill.*	16.2*
Snakes (I) Panama, average of 4 field collections	2672	45	59.4	3.8	9	?	18.2

*Two species both at this level.

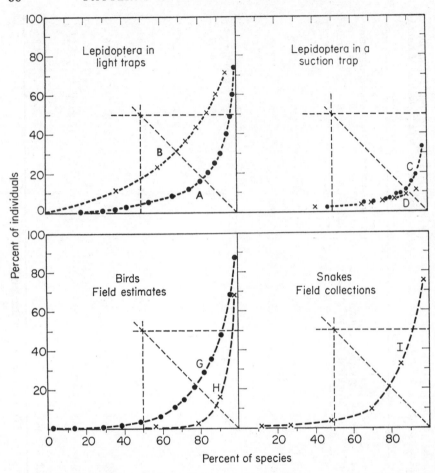

FIG. 23. Diagrams showing the relative proportion of rare and common species in mixed wild populations. The percentage of species (horizontal ordinate) is plotted against the percentage of individuals (vertical ordinate) (see Table 27).

for seven of these, the relation between the accumulated percentage of species as they are arrayed in order of increasing abundance (as the horizontal ordinate) and the corresponding accumulated percentages of individuals (as the vertical ordinate) (see Evans, 1950). In each case we get a curve of the same general form, starting with a rapid increase of species with an almost imperceptible rise in the percentage of individuals. This changes rather suddenly, usually in the neighbourhood of 80 to 90 per cent of the species, to a rapid rise in the number of individuals associated with a relatively slow increase in species.

The curve would be of the same type if the actual accumulated numbers of species and individuals had been used, but by expressing these as percentages the different diagrams are more easily comparable.

In the catch of Macro-Lepidoptera in a light trap at Rothamsted during the year 1935 (Fig. 23A) the 50 per cent rarer species account for only 3.2 per cent of the total individuals, while the single most abundant species alone accounts for 27 per cent. Half of all the individuals belong to the six most abundant species, which is only 3 per cent of the total species represented.

In the much smaller sample of hawk-moths (Sphingidae) caught in a light trap in Nigeria in one night (Fig. 23B, and also p. 36) the insignificance of the rarer species and the dominance of the more abundant ones is less extreme; the 50 per cent rarer species make up 16 per cent of the total individuals and it requires 16 per cent of the commoner species (but, in fact, only three species) to account for 50 per cent of the individuals. The single commonest species accounts for almost the same percentage of the total population as in the previous case, but it must be remembered that with only nineteen species represented, this one species represents 5 per cent of the total species.

The two suction-trap catches of Lepidoptera at Cardington (Table 27 and Fig. 23C, D, see also p. 33) are remarkable in the very high proportion of the total catch determined by the single most abundant species. This, in 1959, amounted to almost 88 per cent of the total individuals. On the other hand, with the Aphidae captured in suction traps at Rothamsted by C. G. Johnson this extreme dominance is not found, the percentage of individuals in the most abundant species being 22.3 per cent, a much more normal figure (Fig. 23G).

In the field counts of birds made by Parmenter (Fig. 23F) (see p. 279) in the London area during the winter, the Starling (*Sturnus vulgaris*) made up just over 50 per cent of all the birds seen, but this may be a disturbance of normal balance due to the habit of this species of congregating in large flocks at this time of the year. The percentage of individuals in the rarer species is very small, and only 0.5 per cent of the population belonged to the 50 per cent rarer species.

In the largest population of all, the 65 million land-nesting birds of England and Wales (Fig. 23H and p. 45), the rarer species are still less significant in the population, only 0.2 per cent of the individuals representing half of the species. At the other end, the two most abundant species (the Chaffinch and the Blackbird *Turdus merula*), were each estimated at 10.5 million individuals or 16 per cent of the population.

An average of four field collections of snakes in Panama (Fig. 23I, see also p. 47) showed only 3.8 per cent of the 2672 individuals belonging to the 50 per cent rarer species, and 50 per cent of the total individuals were in the 9 per cent most abundant species. It is not possible in this case to give the average for the single most abundant species, as the actual species in the different collections are not named.

The form of diagram used in Fig. 23 crowds together the observations in the rarer species. By using a logarithmic scale for the proportion of individuals (but keeping the arithmetic scale for the species) Fig. 24 is obtained. This

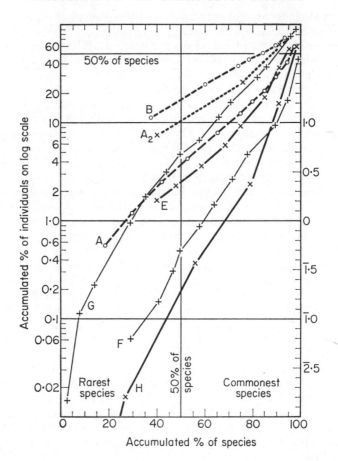

FIG. 24. Relation of accumulated percentage of individuals (on log scale) and accumulated percentage of species, to show relative proportion of the rare and common species. A, Lepidoptera at Rothamsted, year 1935. A_2, as A, for $\frac{1}{8}$ year average. B, Sphingidae in Nigeria one night. E, Aphidae in suction trap at Rothamsted. F, Birds observed near London in winter. G, Birds of Quaker Run Valley. H, British land nesting birds.

tends to straighten out the relation curve, and by using the small fractional value of percentage of individuals (negative values on the log scale) the relationships of the rarer species are much better shown. On the other hand, the higher percentages of individuals are more crowded, and it will be seen that the line of 50 per cent of the individuals is very close to the top of the diagram. In Fig. 24 seven of the samples are indicated as in the legend under the figure. The varying percentage of individuals corresponding to the rarer half of the species is well shown, with the lowest at 0.2 per cent for the 65 million British land-nesting birds (Fig. 24H), and the highest 17 per cent for the single-night catch of hawk-moths in Nigeria, with only sixty-five individuals. In general, the greater the number of individuals in the sample the more the population is dominated by the more abundant species.

EFFECT OF AGGREGATION AND MIGRATION ON SEASONABLY CHANGING POPULATIONS

The problem of the swamping of a local population, such as resident birds in winter, by exceptional aggregations or immigrations of single species is well illustrated by the figures given in the annual records of the Illinois Christmas Bird Census, published in the *Audobon Bulletin*. The number of individuals of one particular species, the Mallard Duck, in relation to the rest of the population for 4 successive years is shown in Table 28.

TABLE 28. *Variation in proportion of the single most abundant species—the Mallard Duck—in the Illinois Christmas Bird Census from 1954 to 1957, showing possible effect of a migrant species on the population balance*

Year	Total species	Total individuals	Mallard duck No.	As % of total
1954	117	201,063	100,000	49.7
1955	115	124,375	35,841	28.8
1956	119	73,008	4,000	5.5
1957	110	581,591	500,000	86.0
All 4 years	146	980,037	639,841	65.3

The variations in numbers of Mallards—from 4000 to 500,000, or from 5.5 to 86 per cent of the population—indicates the difficulties of determining population structure from such records. It is, in fact, unlikely that any regular migrant, even in its breeding area, will give a homogeneous pattern with the resident species. The population levels in such migrants will have been determined largely by conditions of life and competition in another area with a different population complex. A close study of the structure of a bird population over several years at short intervals, including the seasons with and without the regular migrants, would be of considerable interest. After each sudden change of population the previous balance would be upset, and a change would set in towards a new balance, which would not be reached before the next violent seasonal population upset.

It is also probable that the increased amount of food available in temperate climates in the summer may produce conditions of low biological competition producing seasonal changes in the forces leading to population balance.

Chapter 4

PROBLEMS OF SPECIES AND SMALL AREAS, CHIEFLY BOTANICAL

THE information usually available for the study of the structure of plant populations has two important differences from that on which animal studies are based, one leading to greater complexity, the other to greater simplicity.

The first difference arises from the difficulty, in many species of plants with vegetative reproduction, of deciding what is an "individual". This problem does not arise in the case of most animals. The second difference is that, owing to the lack of movement in plants, it is quite easy to say that a certain plant belongs to a particular area, even to a few square inches. With mobile and particularly with winged animals their appearance in a small area is no proof that they are a real part of the local ecological association. Migrant animals form a particularly complicated addition to these studies. As a result of these differences the study of animal populations has been based on numbers of individuals, while in the botanical world the emphasis has been on area.

Thus there has grown up in botanical ecology the use of "quadrats" or small samples of specified area, often a square metre or less, on which only the presence or absence of a species is recorded. The actual number of plants cannot be ascertained. Plant associations have often been classified on the basis of those so-called "dominant" species which occur in a high proportion of the quadrats.

The "quadrat" is—or at least should be—a small random sample of the ecological community under investigation. It contains a probably unknown number of "plant units", but if the quadrats are large enough in relation to the size of the plants, the number of units should be proportional to the number of quadrats. If there are N units in 1 quadrat, we may assume that there are $q N$ in q quadrats.

The plant unit may be considered as that quantity of any plant that is behaving, particularly from the point of view of distribution, as if it corresponded to an individual in an animal population. In a species which produces a definite plant from each seed, and which has no stolons, tillers or other vegetative means of lateral spread, the unit will correspond to the plant. In species which produce compact masses of vegetation all from one original seed, the units may consist of a number of plants, originally and perhaps still in living contact, which behave in such close association as to resemble an individual unit.

It is important to note that in order to be a statistically reliable sample the quadrat must contain a considerable number of units. If some of the plants in the association are large, e.g. trees or large vegetation aggregations, then the quadrats must be increased in size. This is also necessary as the presence of one very large plant in a small quadrat reduces the possible number of plants in the rest of the area.

THE RELATION OF THE NUMBER OF SPECIES TO THE AREA SAMPLED

In the case of animals, where actual numbers of individuals can be counted, it has already been shown that if samples of different sizes are taken from the same population the relation between number of species and the logarithm of the number of individuals can be represented by a line which for small numbers at first rises slowly. The rate of increase becomes greater, but by the time several hundred individuals are represented it becomes constant, giving a straight-line relation over a large range of sample size. If we postulate a log-series distribution this line remains straight up to the limit of the population; if a log-normal or negative binomial is assumed, then at a high value of

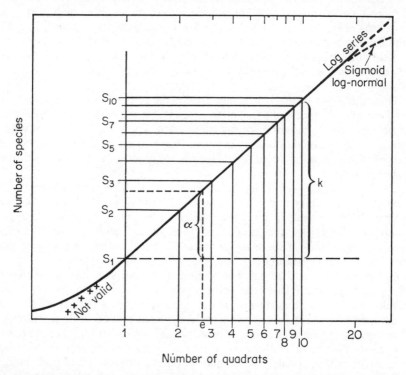

FIG. 25. Relation between the numbers of the species and the numbers of quadrats in the range where $S = k \log Q$. The relation to the index of diversity from the log series is also shown.

N the curve becomes sigmoid with a reduction in the rate of inclusion of new species (see Fig. 25).

Botanists have observed empirically that, over a considerable range of sample size (number of quadrats), but excluding very small areas, there is a straight-line relation between the number of species and the logarithm of the number of quadrats, provided that these are all taken within the limits of one association. Thus within this range it is possible to discuss questions of frequency distribution without committing oneself to any particular mathematical interpretation. This is an initial advantage as it is within this range that most quadrats normally are selected.

A theory, such as the logarithmic series distribution, which deals with the relation over its whole range, will give an estimate of the number of species in a single quadrat if we know the number of units it contains. When we deal only with the central portion of the curve our simplified formula will give the increase in number of species with an increase of the number of quadrats, but not the initial number of species in a single quadrat. This must be found by observation.

The formula has the form

$$S_q - S_1 = k \log q$$

where S_q and S_1 are the number of species in q and in a single quadrat, and k is a constant. It follows from the above that k is the number of species added to those already present in 1 quadrat by a tenfold increase in the number of quadrats; for example, from 1 to 10 or from 10 to 100 quadrats. Since in the previous discussion on animal populations and the log series (p. 21) it has been shown that " α" (which is a measure of diversity, see Chapter 7) is the number of species added by increasing the size of the sample by "e" (2.718). It follows that:

$k = \alpha \log_e 10 = 2.30 \alpha = 2.30 \times$ the Index of Diversity from the logarithmic series.

It also follows that the number of species added at any level by including one extra quadrat is given by

$$S_q - S_{q-1} = k \log_{10} \frac{q}{q-1}.$$

In the case of the logarithmic series, where logs to the base "e" are used, the number of species added by a single quadrat at any level is $\alpha \log_e q/(q-1)$. When q is large $\log_e q/(q-1)$ becomes very close to $1/q$, so the number of species added by the addition of a single quadrat when the previous number is already large is α/q. If we use the constant k and logs to the base 10, $k = 2.30 \alpha$, and $k \log_{10} q/(q-1)$ approximates to $k/2.30 q$, so that the final result is the same pattern.

Figure 25 shows graphically some of the above conceptions. Between 1 and 10 quadrats the relation between species and log quadrats is assumed to be linear. Below 1 quadrat the line cannot be extended in a straight line, but must be curved, otherwise there would be small areas with no species, or even with less than none! Upwards it is not yet certain whether the line remains straight or is curved to a sigmoid form. This will be discussed later. The relation between α and k are also shown. The constant k and logs to the base 10 are easier to use and to understand, but they are not so closely related to mathematical theory.

THE DISTRIBUTION OF SPECIES ON DIFFERENT NUMBERS OF QUADRATS

If we have a number of quadrat samples from a reasonably uniform plant association, each, on an average, will contain the same number of species. Two quadrats together will have a chance of containing more species than 1; 3 quadrats rather more than 2, and so on. The chance of new species, however, gets steadily smaller as the number of quadrats is increased. If p_1 be the number of species found on 1 quadrat, and p_2 the increase by adding a 2nd quadrat, and p_3 the further increase by adding a 3rd quadrat, and so on, it follows that with 2 quadrats there are on an average

$2p_2$ species found on one or other quadrat only,

and $p_1 - p_2$ species found on both quadrats.

With 3 quadrats there are

$3\,p_3$ species found on 1 quadrat only,

$3\,(p_2 - p_3)$ found on any 2 quadrats only,

and $(p_1 - p_2) - (p_2 - p_3)$ species on all 3 quadrats.

For q quadrats the distribution of species is:

on 1 quadrat $\quad = qp_q$

on 2 quadrats only $= \dfrac{q(q - 1)}{1.\,2.} (p_{q-1} - p_q)$

on 3 quadrats only $= \dfrac{q(q - 1)\,(q - 2)}{1.\,2.\,3.} (p_{q-2} - 2p_{q-1} + p_q)$

and the number of species in r out of the q quadrats is

$$^qC_2[p_{q-r+1} - {}^{r-1}C_1 p_{q-r+2} + {}^{r-2}C_2 p_{q-r+3} - \ldots\ldots + (-1)^{r-1}p_q].$$

The calculations of these values can be simplified by putting the number of species in 1, 2, 3, etc., quadrats (S_1, S_2, S_3, etc.) in a vertical column, as in Table 29: the differences between these, p_1, p_2, p_3, etc., in a second column with p_1 opposite S_1 and then again the differences in a third column with $p_1 - p_2$ opposite to p_2; and so on. Then any line read horizontally will give, in the successive differences from p_n onwards, the number of species in two particular quadrats, the number in any three particular quadrats, etc.

To obtain the total number of species in all possible single quadrats, in all possible pairs, in all possible threes, etc., the column differences must be multiplied by ^qC_r as shown in heavy type numbers in the same table. This is independent of any particular theory of the relative abundance of species.

The table also shows a numerical example based on samples taken from a population based on a logarithmic series with a Diversity (α) of 10 and 100 units in each quadrat.

It has been shown earlier in this chapter that for a large number of quadrats, the difference between the number of species in q and $q - 1$ quadrats approaches α/q. If we insert in the general formula on p. 68 the value of α/q for p_q, $\alpha/q - 1$ for p_{q-1}, etc., we get:

$$\text{the number of species in 1 quadrat} = qp_1 = q\left(\frac{\alpha}{q}\right) = \alpha,$$

$$\text{the number in 2 quadrats} = \frac{q(q-1)}{1.\,2.}\left[\frac{\alpha}{q-1} - \frac{\alpha}{q}\right] = \frac{\alpha}{2},$$

the number in 3 quadrats =

$$\frac{q(q-1)(q-2)}{1.\,2.\,3.}\left[\frac{\alpha}{q-2} - \frac{2\alpha}{q-1} + \frac{\alpha}{q}\right] = \frac{\alpha}{3}.$$

So the distribution of the number of species in 1, 2, 3 or more quadrats out of a large number approaches to a hyperbolic series with the first term equal to α, the index of diversity of the population. This, however, holds only for the earlier terms. For the later terms—species which occur in a large proportion of the quadrats—the values are above the hyperbola, and for large quadrats there is a rapid increase in the number of species found in all or nearly all the quadrats (see Figs. 26 and 27).

TABLE 29. *Method of differences used in calculating the expected numbers of species to be found in 1 quadrat only, in 2 quadrats, in 3 quadrats, and so on*

Above, general theory; below, numerical example (based on the logarithmic series, where $\alpha=10$ and $N=100$). The figures in any one column are the differences between two figures in the previous column.

No. of quadrats	Total no. of species	First difference. Species in 1 quadrat only	Second difference. Species in 2 quadrats only	Third difference. Species in 3 quadrats only	Fourth difference. Species in 4 quadrats only
1	S_1	$1 \times p_1$	—	—	—
2	S_2	$2 \times p_2$	$p - p_2$	—	—
3	S_3	$3 \times p_3$	$3 \times (p_2-p_3)$	$(p_1-p_2) - (p_2-p_3)$	—
4	S_4	$4 \times p_4$	$6 \times (p_3-p_4)$	$4 (p_2-p_3) - (p_3-p_4)$	difference
5	S_5	$5 \times p_5$	$10 \times (p_4-p_5)$	$10 (p_3-p_4) - (p_4-p_5)$	$5 \times$ difference
6	S_6	$6 \times p_6$	$15 \times (p_5-p_6)$	$20 (p_4-p_5) - (p_5-p_6)$	$15 \times$ difference, etc
1	23.98	23.98	—	—	—
2	30.44	2×6.47	17.51	—	—
3	34.34	3×3.89	3×2.57	14.94	—
4	37.14	4×2.80	6×1.10	4×1.47	13.47
5	39.32	5×2.18	10×0.61	10×0.49	5×0.99
6	41.11	6×1.79	15×0.39	20×0.22	15×0.26 etc.

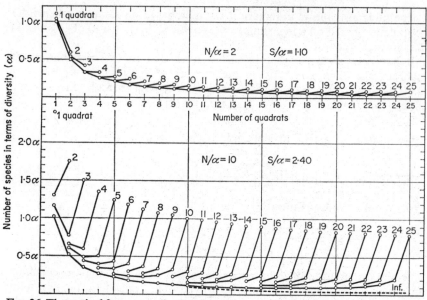

FIG. 26. Theoretical frequency distributions of species (in terms of α) in 1–25 quadrats for values of $N/\alpha = 2$ and $N/\alpha = 10$ or $S/\alpha = 1.1$ and 2.4. (From *J. Anim. Ecol.* (1950), **38**, 114.)

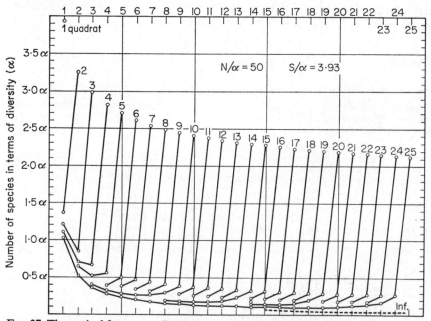

FIG. 27. Theoretical frequency distribution of species (in terms of α) in 1–25 quadrats for value of $N/\alpha = 50$ or $S/\alpha = 3.93$. (From *J. Anim. Ecol.* (1950), **38**, 115.)

In the log series we have shown that

$$S = \left(\log_e \left(1 + \frac{N}{\alpha} \right) \right)$$

where N is the number of individuals in the sample, and α the index of diversity. This can be written as

$$S = k \log_{10} \left(1 + \frac{2.3\,N}{k} \right)$$

or for a large sample $S = k \log_{10} \left(\frac{2.3\,N}{k} \right)$.

If we call the assumed number of units in a quadrat N_1, then the number of species in a group of q quadrats is

$$S_q = k \log_{10} \left(\frac{2.3\,q\,N_1}{k} \right) = k \left[\log_{10} q + \log_{10}\frac{N_1}{k} + \log_{11} 2.3 \right].$$

So the number of species in a group of quadrats depends on three variable factors as follows:

(1) directly on k, which is a measure of diversity and is a constant for any particular population;

(2) on the log of the number of quadrats, which can be varied by the observer;

(3) on the log of the ratio between N_1 and k which for any given value of k can be varied by altering the size of the quadrats.

Thus by choosing quadrat sizes in different populations in a constant ratio to the diversity, the different types of distribution would depend only on the value of the diversity.

Again, basing the argument on the log series 9 (which not only fits the straight-line portion of the relation between log-sample size and number of species, but also the short lower curved portion) any particular value of N/α gives a single possible value for "x" (since $N/\alpha = x/(1 - x)$), and hence single values of S/α and of S/N which is the average number of individuals per species. So, if we wish the quadrats in two different associations to have the same ratio of N/α, the same constancy must be found in S/α and in N/S. As it is much easier to count the species in a quadrat than to estimate the number of individuals, the use of S/α, or $2.3\,S/k$, is of greater simplicity and more practical value in the field. It must be remembered, however, that changes in quadrat size are *directly* proportional to changes in N/α, but not so to changes in S/α.

The relations between the values of these different expressions for different values of N/α are given in Table 30. With these in view, Table 31 and Figs. 26 and 27 have been prepared to show the theoretical frequency of species in different numbers of quadrats up to 25, when N/α is 2, 10, or 50, or when S/α is 1.1, 2.4, or 3.9.

In the figures the vertical scale is in terms of α, so they must be multiplied by α, or by $k/2.30$, to get the actual number of species.

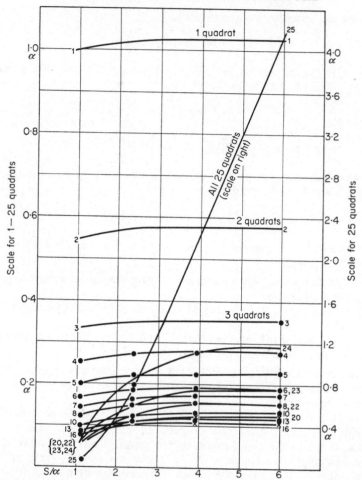

Fig. 28. Diagram showing the number of species (in terms of α) found on any number of quadrats out of 25, for any value of S/α from 1–6. Calculated values at 1.1, 2.4, 3.93, and 5.99, the rest by interpolation. Scale for 1–24 quadrats on the left; for 25 quadrats on the right. (From *J. Anim. Ecol.* (1950), **38**, 116.)

TABLE 30. *Relation between the number of species (S), the number of individuals (N), the diversity (α), and the value of x, in populations based on a log-series frequency distribution*

N/α	x	S/α	N/S
2	.6666	1.10	1.82
10	.9091	2.40	4.17
50	.9804	3.93	12.72
100	.9901	4.62	21.67
400	.9975	5.99	66.73
1000	.9990	6.91	144.7

TABLE 31. *Frequency distribution (in terms of α) of species in various numbers of quadrats up to 25, with different values of the ratios of N/α and S/α. Note that, with increasing number of quadrats, the number of species found in only 1 quadrat approaches the value of α*

No. of quad-rats	Total no. of species	1	2	3	4	5	6	7	8	9	10	11	12	13	14	15	16	17	18	19	20	21	22	23	24	25
													$N/\alpha=2$: $S/\alpha=1.10$													
1	1.098	1.098																								
5	2.398	1.003	0.506	0.345	0.273	0.270																				
10	3.044	1.001	0.501	0.335	0.253	0.204	0.173	0.151	0.138	0.133	0.155															
15	3.434	1.000	0.501	0.334	0.251	0.202	0.169	0.145	0.128	0.115	0.105	0.097	0.092	0.089	0.091	0.114										
20	3.714	1.000	0.500	0.334	0.251	0.201	0.168	0.144	0.126	0.113	0.102	0.093	0.086	0.080	0.076	0.072	0.069	0.068	0.068	0.071	0.092					
25	3.932	1.000	0.500	0.334	0.250	0.200	0.167	0.144	0.126	0.112	0.101	0.092	0.085	0.079	0.073	0.069	0.065	0.062	0.059	0.057	0.055	0.054	0.054	0.055	0.059	0.078
													$N/\alpha=10$: $S/\alpha=2.40$													
1	2.398	2.398																								
5	3.932	1.091	0.613	0.485	0.494	1.248																				
10	4.615	1.043	0.547	0.385	0.308	0.267	0.246	0.241	0.255	0.318	1.005															
15	5.172	1.028	0.529	0.365	0.284	0.236	0.206	0.186	0.173	0.165	0.161	0.163	0.172	0.195	0.259	0.896										
20	5.303	1.021	0.522	0.356	0.274	0.225	0.193	0.171	0.155	0.143	0.134	0.128	0.124	0.122	0.121	0.124	0.130	0.142	0.166	0.227	0.829					
25	5.525	1.016	0.517	0.351	0.268	0.219	0.186	0.163	0.147	0.134	0.124	0.116	0.110	0.105	0.102	0.099	0.098	0.097	0.098	0.100	0.104	0.111	0.124	0.148	0.206	0.782
													$N/\alpha=50$: $S/\alpha=3.93$													
1	3.932	3.932																								
5	5.525	1.111	0.639	0.523	0.564	2.690																				
10	6.217	1.051	0.557	0.396	0.321	0.282	0.265	0.265	0.291	0.386	2.403															
15	6.621	1.033	0.536	0.371	0.291	0.244	0.215	0.195	0.183	0.177	0.176	0.181	0.196	0.230	0.325	2.267										
20	6.909	1.025	0.526	0.361	0.279	0.230	0.198	0.177	0.161	0.150	0.142	0.136	0.133	0.132	0.134	0.139	0.148	0.165	0.200	0.292	2.182					
25	7.132	1.020	0.520	0.355	0.272	0.223	0.191	0.168	0.151	0.139	0.129	0.121	0.116	0.111	0.108	0.107	0.106	0.106	0.108	0.112	0.119	0.129	0.147	0.181	0.270	2.121

Figure 28 shows the theoretical distribution of species in any number of quadrats up to 25, made from calculated values for $S/\alpha = 1.1, 2.4, 3.93$ and 5.99, and other values by interpolation. It will be noticed that once the sample is large (i.e. $S/\alpha = 3$ or more) the ratio makes little difference to the number of species in quadrats up to nearly all of the total, but a very great difference to the number found in all the quadrats, whatever their number may be.

It follows from the above that if we have two plant associations, one richer than the other, they will give relatively different frequency distributions of species in 1, 2, 3, etc., quadrats unless the size of the quadrat, in units or numbers of species, has been chosen in relation to the diversity of each.

For example, it will be seen from the tables that if, within an association, 5 quadrats are taken with a size corresponding to $N/\alpha = 10$, or $S/\alpha = 2.4$, there will be 1.09 α species in 1 quadrat only, 0.61 α species in 2 quadrats, 0.49 α in 3, 0.49 α in 4 and 1.25 α species in all 5 quadrats. If, however, the quadrats had been chosen five times as large ($N = 50\alpha$) the numbers of species would have been 1.11α, 0.64α, 0.52α, 0.56α and 2.70α respectively. The increase in size of the quadrat has increased the number of species common to all quadrats.

It is also important to note that the number of species which occur in x quadrats out of y is not the same as the number which occur in $2x$ quadrats out of $2y$. This mistake has been made by several botanists when they were unable to get the full number of quadrats that they wished in one particular association.

THE NUMBER OF SPECIES FOUND IN DIFFERENT PERCENTAGES OF THE QUADRATS

Certain botanists, and particularly Raunkaier (1934), have developed a simple convention of grouping together the species found on a number of quadrats into five divisions as follows:

I. Species found in only 1–20 per cent of the quadrats, the rarer species.
II. Species found on 21–40 per cent of the quadrats, not so rare.
III. Species found on 41–60 per cent of the quadrats.
IV. Species found on 61–80 per cent of the quadrats.
V. Species found on 81–100 per cent of the quadrats.

The species in Group V are the common or "dominant" species, which are usually considered as characteristic of the association which is being studied.

It is interesting to apply the theories just discussed to see how far the distribution of species in these five groups simplifies the analysis, how far it gives a reliable measure of some character of the population pattern, and, on the other hand, how far it may depend on such artificial factors as the size and number of the quadrats and the total area sampled.

The information in Table 32 and Fig. 29 has been extracted from the tables already given to show the theoretical numbers of species expected to occur in Raunkaier's five groups, in the case of 5, 10, 15, 20 and 25 quadrats,

TABLE 32. *The number of species, in terms of α and as a percentage of the total species, in Raunkaier's five groups of quadrats; i.e. species in 1–20, 21–40, 41–60, 61–80, and 81–100 per cent of the total quadrats*

No. of quadrats	Total no. of species	Group I		Group II		Group III		Group IV		Group V	
		No.	%	No.	%	No.	%	No.	%	No.	%
		$N/α = 2:S/α = 1.1$									
5	2.40α	1.00α	42	0.51α	21	0.35α	14	0.27α	11	0.27α	11
10	3.04	1.50	49	0.59	19	0.38	12	0.29	9	0.29	9
15	3.43	1.84	53	0.62	18	0.39	11	0.29	9	0.29	9
20	3.71	2.09	56	0.64	17	0.39	11	0.30	8	0.30	8
25	3.93	2.28	58	0.65	17	0.40	10	0.30	8	0.30	8
		$N/α = 10:S/α = 2.4$									
5	3.93α	1.09	28	0.61	16	0.49	12	0.49	13	1.25	32
10	4.62	1.59	34	0.69	15	0.51	11	0.50	11	1.32	29
15	5.02	1.92	38	0.73	14.5	0.52	10	0.50	10	1.35	27
20	5.30	2.17	41	0.74	14	0.53	10	0.50	9	1.36	26
25	5.53	2.37	43	0.75	14	0.53	10	0.50	9	1.37	25
		$N/α = 50:S/α = 3.93$									
5	5.53α	1.11	20	0.64	12	0.52	9	0.56	10	2.69	49
10	6.22	1.61	26	0.72	12	0.55	9	0.56	9	2.79	45
15	6.62	1.94	29	0.75	11	0.56	8	0.55	8	2.82	43
20	6.91	2.19	32	0.77	11	0.56	8	0.55	8	2.84	41
25	7.13	2.39	34	0.78	11	0.56	8	0.55	8	2.84	40
		$N/α = 400:S/α = 5.99$									
5	7.60α	1.12	15	0.64	8	0.53	7	0.58	8	4.73	62
10	8.29	1.61	19	0.72	9	0.56	7	0.57	7	4.83	58
15	8.70	1.95	22	0.75	9	0.56	6	0.57	7	4.87	56
20	8.99	2.20	24	0.77	9	0.57	6	0.57	6	4.89	54
25	9.21	2.39	26	0.78	9	0.57	6	0.57	6	4.90	53

and with quadrats of different sizes representing different ratios of N/S. All numbers of species are in terms of α, the diversity of the population.

If we consider in Table 31 the data for 25 quadrats we see that as the quadrat size increases from $N = 2α$ to $N = 400α$ there is very little alteration in the number of species in Group I (1–5 quadrats) from 2.28 to 2.29α only; but there is a rapid fall in the percentage of the total number of species from 58 to 26 per cent. The changes in numbers in Groups II to IV are also small. In Group V, on the contrary, there is a large increase in both actual number of species (0.30–4.90)α and in percentage from 8–53 per cent). In fact, of the total increase of 5.28α species, due to the increase in quadrat size, 4.60α are added to Group V. As the quadrat size has increased 200 times, many of the original species in Group I have moved into higher groups. It is obvious that if the quadrats had been very large, say an acre or a hectare, most of the species will be found in all or nearly all of the quadrats.

The effect of increasing the number of quadrats without altering their size

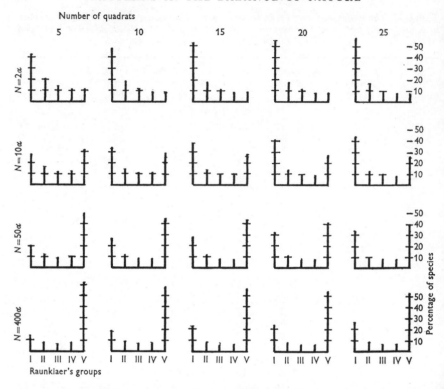

FIG. 29. Theoretical percentages of total species which occur in Raunkaier's five groups of quadrats, according to variations in the size and number of the quadrats. (From *J. Anim. Ecol.* (1950), **38**, 118.)

has the reverse effect. It will be seen from the table and figure that, at all levels of quadrat size considered, an increase in the number of quadrats produces an increase in both number and percentage in the rarer species in Group I. In the higher groups there is always an increase, though sometimes very small, in the number of species, but owing to the increased total, the percentage may fall. In Group V the fall in percentage may be considerable. Thus when $N = 2\alpha$ out of the 1.53 species added by the increase in the number of quadrats from 5 to 25, 1.28α are added to Group I and only 0.17α to Group V, resulting in a fall of percentage from 62 to 53 in the latter.

It is difficult to calculate the theoretical distribution of s species in a number of quadrats greater than 25. Figure 30, prepared from the information for $N = 10\alpha$ in Table 30, shows that there is an almost straight-line relation between the percentage of the total species in each of Raunkaier's groups and the log of the number of quadrats up to 25. These lines have therefore been extended to show the probable percentage of species in the groups up to a total of 100 quadrats. For example, when $N = 10\alpha$, there should be a total of

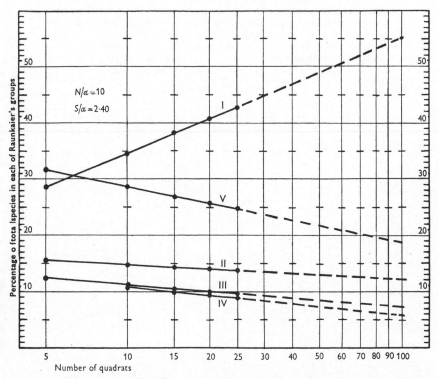

FIG. 30. The change in the percentages of total species found in each of Raunkaier's five groups as the number of quadrats is increased for the same association in which $N/\alpha = 10$. Up to 25 quadrats by calculation, above 25 by extrapolation. (From *J. Anim. Ecol.* (1950), **38**, 119.)

6.90α species, and the percentage of these in the five groups would be approximately 55, 12, 7, 6 and 19.

If we alter both the number and size of the quadrats in an inverse ratio, thus keeping the same total area sampled, we again get inconsistent results as shown in Table 33 and Fig. 31. It will be seen that in each example the larger number of small quadrats gives a higher proportion of the "rarer" species in Group I; and a smaller number of larger quadrats gives more dominant species in Group V.

So we see that the number and proportion of species in these five groups of quadrats depend very largely on the size and number of the quadrats. The effect of quadrat size can be partly eliminated by choosing it in relation to the diversity of the population. Since, however, it is probable that this characteristic is what determines the fundamental pattern of distribution of species in the quadrats (see p. 71), if we can find the diversity by other simpler means we do not need the elaborate and uncertain technique just discussed.

TABLE 33. *Four samples of the same total area divided into more small or fewer large quadrats, illustrating the effect on the percentage of species in each of Raunkaier's groups*

Sample and total sp.	Quadrats No.	Size ($N/\alpha =$)	Raunkaier's groups (% of species) I	II	III	IV	V
A. 3.94α	10	5	39.8	17.0	11.9	10.9	20.3
	5	10	27.7	15.5	12.4	12.4	31.7
B. 4.61α	10	10	34.5	15.0	11.1	10.8	28.6
	5	20	23.9	13.7	11.1	11.5	40.0
C. 6.22α	10	50	25.9	11.6	8.8	9.0	44.9
	5	100	17.8	10.3	8.5	9.2	54.0
D. 5.50α	25	10	43.1	13.6	9.6	9.1	24.9
	5	50	20.2	11.6	9.5	10.2	48.9

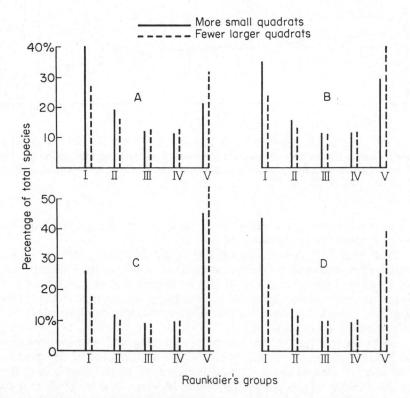

FIG. 31. Effect on the percentage of species in Raunkaier's five quadrat groups, of dividing an area into a larger number of small quadrats or a smaller number of large quadrats. For details, see Table 33.

JACCARD'S OBSERVATIONS IN SWITZERLAND

For comparison with the theoretical suggestions just discussed, information is required from field observations on the presence or absence of species in a number of quadrats.

In 1908, P. Jaccard discussed the distribution of ninety-two species of plants on 52 quadrats, each measuring 1m², in the Alpine valley of the Grand Eau near Ormont, Vaud, Switzerland, at a height of about 1200m (4000 ft). The quadrats were taken, in groups of 2 to 8, in nine different areas, in each case the quadrats being in contact with one another on at least one side. The different areas were about a kilometre apart up the valley, except for IV and V which were in the same meadow.

Figure 32 has been rearranged from Jaccard's data so that the species are in order of frequency of occurrence in quadrats. The names of the species in Table 34 are as given by Jaccard and no attempt has been made to bring the nomenclature up to date, as our concern here is not with the actual identity of any of the species, but with the pattern as a whole.

Table 35 shows the number of species in 1, 2, 4 and all quadrats in each of Jaccard's nine areas, and Fig. 33 shows the same information in a graphical form plotted as number of species against log area or number of quadrats. It will be seen that (excluding IX, which has only 2 quadrats) the five areas II, III, IV, VI and VII all show an almost straight-line relation between log area and number of species as expected from the theories just discussed. In the other three, I, V, and VIII, the relation is slightly curved, but even here the departure from the straight line is usually not more than two species. Such departures would be expected if species which would have normally occurred in most of the quadrats were prevented from surviving in some by a lack of uniformity in the area sampled.

Since the slope of the line is a measure of the diversity of the population (see Chapter 7), it would appear that Area I was the most diverse and VII, or possibly IX, the least. In Table 35 are shown the diversities calculated on the assumption of the applicability of the log series. They range from 12.1 for the first area to 3.4 and 3.6 for Areas VII and IX. If the constant k is used these values must be multiplied by 2.3.

Table 36 shows the average number of species in each of Jaccard's areas which occurred in 1, 2, 3, etc., quadrats, together with the expected number calculated from the theory. The two sets of values are also shown in Fig. 34, the observed data as vertical columns of dots, and the estimates by the lighter broken line. It will be seen, particularly from the figure, that there is in general a good resemblance between the observed and the calculated results. The fit is particularly close in Areas V to IX, especially at the two ends, with the species in only one or in all the quadrats. Area IV is good at each end, but irregular in the middle quadrats. The poorest fit is in Areas I and III, which have the largest number of quadrats. Under field conditions the larger the number of quadrats the less likely are they to resemble theoretical samples from a uniform area. The generally too low number of species observed in all

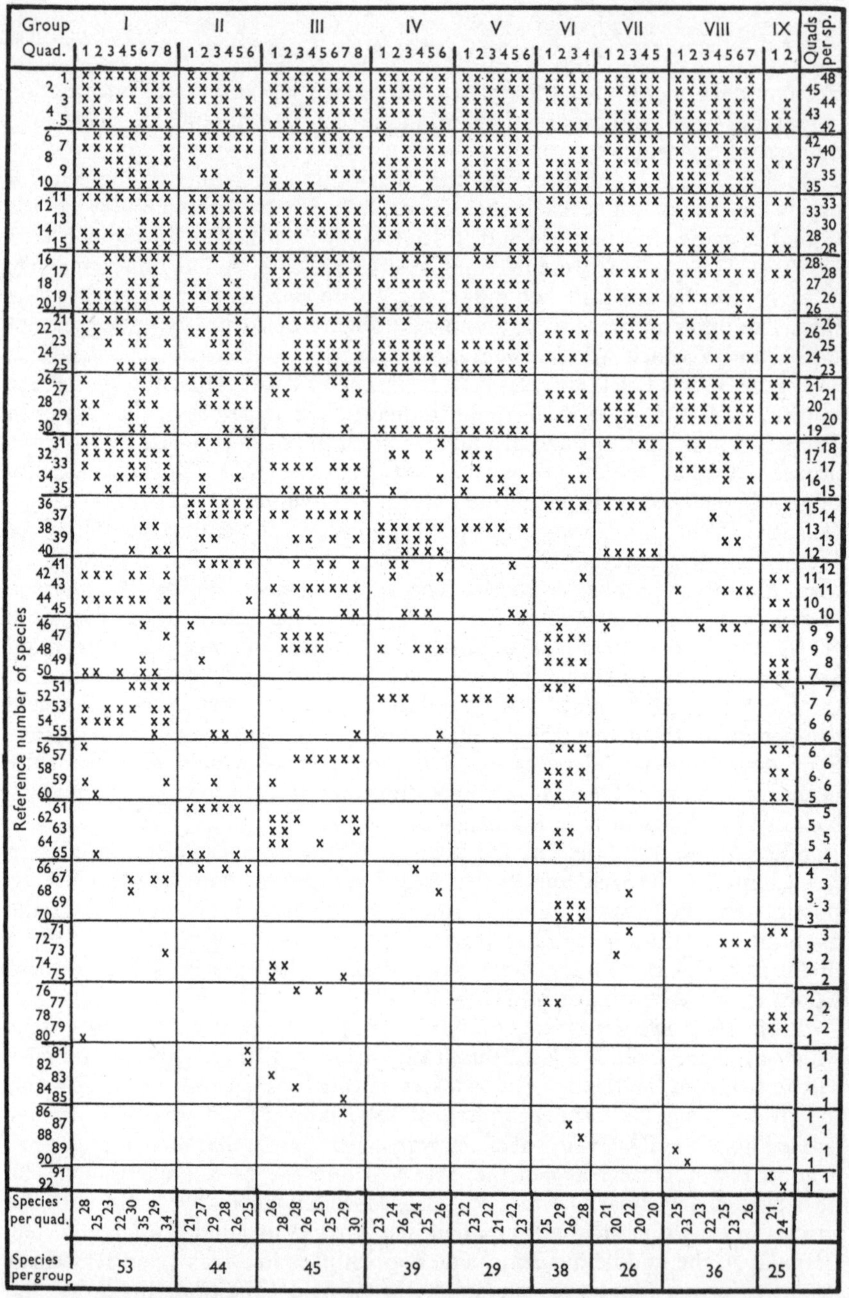

Group	I	II	III	IV	V	VI	VII	VIII	IX	Quads. per sp.
Quad.	1 2 3 4 5 6 7 8	1 2 3 4 5 6	1 2 3 4 5 6 7 8	1 2 3 4 5 6	1 2 3 4 5 6	1 2 3 4	1 2 3 4 5	1 2 3 4 5 6 7	1 2	

(matrix of species occurrences by reference number of species, rows 1–92)

| Species per quad. | 28 25 23 22 30 35 29 34 | 21 27 29 28 26 25 | 26 28 28 25 25 29 30 | 23 26 24 25 26 | 23 22 21 22 23 | 25 29 26 28 | 20 21 20 20 | 25 23 22 25 23 26 | 21 24 | |
| Species per group | 53 | 44 | 45 | 39 | 29 | 38 | 26 | 36 | 25 | |

FIG. 32. The distribution of ninety-two species of plants in 52 quadrats of 1 m², as observed by Jaccard in an Alpine Valley in Switzerland (redrawn from Jaccard, 1908). (From *J. Anim. Ecol.* (1950), **38**, 124.) (For names of species, see Table 34.)

TABLE 34. *List of species of plants in Jaccard's 52 quadrats in Switzerland, rearranged in order of frequency of occurrences in quadrats, with the number of occurrences in brackets*

1 *Trifolium pratense* (48)
2 *Alchemilla pratensis* (45)
3 *Chrysanthemum leucanthemum* (44)
4 *Festuca pratensis* (43)
5 *Leontodon hispidus* (42)
6 *Dactylis glomerata* (42)
7 *Campanula romboidalis* (40)
8 *Taraxacum officinale* (37)
9 *Lathyrus pratensis* (35)
10 *Ranunculus acer* (35)
11 *Colichicum autumnalis* (33)
12 *Trisetum flavescens* (33)
13 *Geranium silvaticum* (30)
14 *Brunella vulgaris* (28)
15 *Avena pubescens* (28)
16 *Anthoxanthum odoratum* (28)
17 *Vicia cracca* (28)
18 *Anthriscus sylvestris* (27)
19 *Alectorolophus hirsutus* (26)
20 *Cynosurus cristatus* (26)
21 *Poa pratensis* (26)
22 *Carum carvi* (26)
23 *Rumex acetosa* (25)
24 *Trifolium repens* (24)
25 *Polygonium bistorta* (23)
26 *Sanguisorba minor* (21)
27 *Lotus corniculatus* (21)
28 *Medicago lupulina* (20)
29 *Plantage media* (20)
30 *Crepis taraxifolia* (19)
31 *Briza media* (18)
32 *Plantago lanceolata* (17)
33 *Cerastium caespitosum* (17)
34 *Tragapogon pratense* (16)
35 *Phyteuma spicatum* (15)
36 *Silene inflata* (15)
37 *Knautia arvensis* (14)
38 *Melandrum silvestre* (13)
39 *Bromus erectus* (13)
40 *Agrostis canina* (12)
41 *Myosotis silvatica* (12)
42 *Primula elatior* (11)
43 *Deschampsia caespitosa* (11)
44 *Carex sempervivens* (10)
45 *Vicia sepium* (10)
46 *Centaurea jacea* (9)

47 *Festuca ovina* (9)
48 *Achillea millefolium* (9)
49 *Alectorolophus minor* (8)
50 *Linum catharticum* (7)
51 *Campanula rotundifolia* (7)
52 *Veronica chamaedrys* (7)
53 *Bellis perennis* (6)
54 *Equisetum avense* (6)
55 *Galium asperum* (6)
56 *Anthyllis vulneria* (6)
57 *Ranunculus montanus* (6)
58 *Gentiana campestris* (6)
59 *Thymus serphyllum* (6)
60 *Polygala vulgaris* (5)
61 *Ranunculus bulbosus* (5)
62 *Trifolium badium* (5)
63 *Carex pallescens* (5)
64 *Potentilla silvestris* (5)
65 *Viola tricolor* (4)
66 *Centaurea montana* (4)
67 *Geum urbanum* (3)
68 *Ranunculus aconitifolius* (3)
69 *Ajuga reptans* (3)
70 *Picia excelsa* (3)
71 *Phyteuma orbiculaire* (3)
72 *Polygonium aviculare* (3)
73 *Cirsium oleraceum* (2)
74 *Crepis paludosa* (2)
75 *Luzula silvation* (2)
76 *Hieraceum pilosella* (2)
77 *Gentiana verna* (2)
78 *Carlina acaulis* (2)
79 *Hippocrepis comosa* (2)
80 *Viola hirta* (1)
81 *Agropyrum repens* (1)
82 *Hypericum quadrangulum* (1)
83 *Aposeris foetida* (1)
84 *Trollius europaeus* (1)
85 *Astransia major* (1)
86 *Juncus lamprocarpus* (1)
87 *Veronica teucrium* (1)
88 *Crepis aurea* (1)
89 *Listera ovata* (1)
90 *Veratum album* (1)
91 *Brachypodium pinnatum* (1)
92 *Biscutella laevigata* (1)

TABLE 35. *Number of species of plants in different number of quadrats in Jaccard's seven habitat groups in Switzerland, together with calculations of Diversities based on the log series*

Average no. of sp.	Jaccard's group nos.								
	I	II	III	IV	V	VI	VII	VIII	IX
In 1 quadrat	28.3	26.0	27.1	24.7	22.2	27.0	20.6	23.9	22.5
In 2 quadrats	38.4	33.5	34.6	30.8	25.6	32.5	23.3	29.1	25.0
In 4 quadrats	46.2	—	40.3	—	—	38.0	—	33.4	—
In all quadrats	53	44	45	39	29	—	26	36	—
No. of quadrats	8	6	8	6	6	4	5	7	2
Index of Diversity (α)	12.1	10.1	8.7	8.0	3.8	8.0	3.4	6.3	3.6

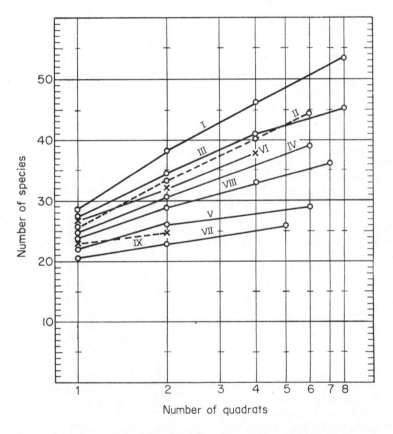

FIG. 33. The relation between the number of species and log area (number of quadrats) in the nine groups into which Jaccard's 52 quadrats were divided (see Fig. 32 and Table 35).

quadrats would be expected from any lack of uniformity in the terrain. The small deficits in the number of species in one quadrat only might well be due to a few of the rarest species occasionally being missed in the field.

The theoretical distribution of species in Raunkaier's five groups can be partly tested from Jaccard's observations over the whole area. Of the ninety-two species that he found in the 52 quadrats, forty-nine occurred in 10 or less quadrats, which corresponds to Raunkaier's first group. The diversity of the whole 52 quadrats is 17, so that there were 2.88α species in Group I. On the theory the first 10 out of 50 quadrats should be very close to a hyperbola with the first term equal to α. Summing the hyperbola there should be 2.93α species, which is remarkably close to the observed value. Since the area included was far from uniform, close agreement would not be expected in the higher groups.

TABLE 36. *The number of species of plants in each of Jaccard's nine groups, which occur in different numbers of quadrats, together with the corresponding numbers calculated from the log series*

Jaccard's group no.		No. of sp. which occur only in quadrats							Total no. of species in all quadrats
		1	2	3	4	5	6	7	8	
I	Obs.	11	4	4	8	6	11	5	4	53
	Cal.	12.6	6.8	4.9	4.0	3.7	3.7	4.4	13.0	
II	Obs.	11	3	8	5	7	10	—	—	44
	Cal.	10.9	6.0	4.5	4.1	4.5	13.2	—	—	
III	Obs.	7	4	3	7	2	5	11	6	45
	Cal.	9.2	5.0	3.6	3.0	2.8	2.9	3.5	15.1	
IV	Obs.	8	3	7	4	5	12	—	—	39
	Cal.	8.7	4.8	3.7	3.4	3.8	14.4	—	—	
V	Obs.	3	1	4	3	4	14	—	—	29
	Cal.	4.1	2.1	1.8	1.7	2.0	17.1	—	—	
VI	Obs.	8	8	4	18	—	—	—	—	38
	Cal.	9.1	5.5	5.1	18.0	—	—	—	—	
VII	Obs.	4	0	3	5	14	—	—	—	26
	Cal.	3.8	2.2	1.8	2.0	16.1	—	—	—	
VIII	Obs.	4	6	2	3	4	6	11	—	36
	Cal.	6.8	3.7	3.0	2.3	2.4	2.9	15.1	—	
IX	Obs.	5	20	—	—	—	—	—	—	25
	Cal.	5.6	19.6	—	—	—	—	—	—	

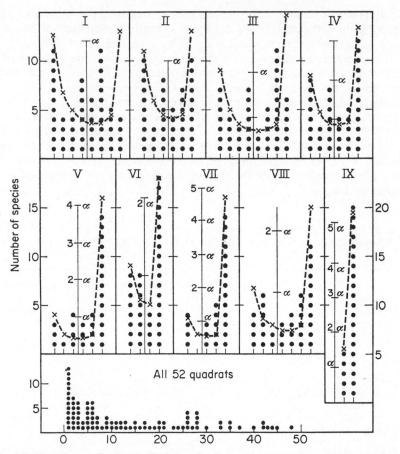

FIG. 34. Frequency distribution of species in quadrats in nine groups of plant associations in Switzerland according to Jaccard. The dots represent the observed data and the broken lines calculated values based on the logarithmic series.

OBSERVATIONS IN NEW SOUTH WALES BY PIDGEON AND ASHBY

In 1940, Pidgeon and Ashby made a survey of some arid land near Broken Hill, New South Wales, to estimate the value of fencing in protecting land from the erosion effects of grazing by stock and by rabbits. Owing to the sparse vegetation it was possible on a number of quadrats to count the actual number of individual plants for each species.

Observations were made on four areas defined as (1) West Reserve, fenced, (2) West Reserve, unfenced, (3) South Reserve, fenced, and (4) South Reserve, unfenced. In addition (5) a large permanent quadrat, of 50 metres square, was studied in one of the fenced areas. Where small quadrats were used they were of elongate form, 10 m by 15 cm, with an area of 1.5 m². Fifty contiguous quadrats were examined in each of the Areas 1 and 2; forty in Area 3, and 39 in Area 4.

For Areas 1, 2, 3 and 5 the number of species in blocks of different sizes up to 200 m² (but not from the quadrats) was counted, and also for the whole of each area, as given in Table 37. Figure 35 shows the same information

TABLE 37. *Numbers of species of plants in different areas at Broken Hill, New South Wales*

Area (sq m)	West fenced	West unfenced	South fenced	Permanent quadrat
1	12	8	5	5
5	18	16	12	9
10	19	17	13	10
50	23	24	17	17
100	28	26	21	21
200	31	29	27	24
12,000 (3 acres)	—	37	—	—
32,400 (7 acres)	50	—	47	—
α cal from 1 and 200 m₂	4.0	3.6	4.2	3.6

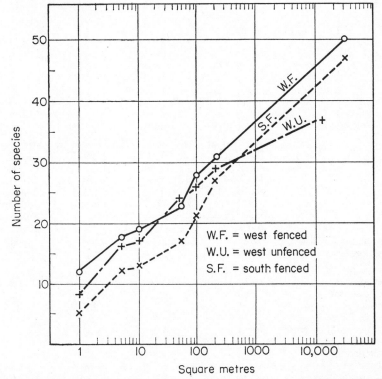

FIG. 35. Number of species of plants in relation to total area of quadrats in different plots of arid land in New South Wales studied by Pidgeon and Ashby.

graphically plotted against a log scale of area. The results are somewhat irregular, perhaps owing to there being only one count at each level. They are, however, except for the total species on the 3 acres in West unfenced, seldom more than two species out of the straight-line relation. The table also gives estimates of the diversity from the increase of species between 1 and 200 m².

Table 38 gives the information from the study of the quadrats in the four areas, including the number of species with different numbers of individuals. In the original publication the numbers of individuals had been multiplied up to bring them to a standard of 100 m². As this is not desirable without some corresponding compensation in the number of species, I have converted them

TABLE 38. *Number of species of plants with different numbers of individuals in four areas of arid land in New South Wales*

No. of individuals per species	West fenced, 50 quadrats, 75 m²	West unfenced, 50 quadrats, 75 m²	South fenced, 40 quadrats, 60 m²	South unfenced, 39 quadrats, 58·5 m²
1	7	6	9	4
2	0	3	5	3
3	6	3	2	2
4	0	0	0	0
5	3 (16)	4 (16)	0 (16)	0 (9)
6	1	0	0	3
7	0	1	1	1
8	0	1	1	2
9	0	1	3	1
10	1 (2)	0 (3)	0 (5)	1 (8)
	And also at 12, 17 (2), 23, 35, 38, 44 (2), 48, 98, 99, 118, 279, 316, 335, 719 and 1849	And also at 16, 17, 45, 48, 73, 98, 131, 154, 224, 758 and 1155	And also at 12, 20, 22, 24, 27 (2), 44, 66, 110, 131, 234, 346 and 667	And also at 11, 12, 13 (3), 15, 23, 34, 51, 107, 274, 337, 366, 460 and 698
Total plants (qN)	4147	2784	1797	2503
Total species S	35	30	34	32
Average plants:				
Per species	118	93	51	78
Per quadrat	83	56	45	64
Per sq m	55	37	30	43
Index of Diversity	5.3	4.8	6.2	5.2
α from qN and S_q from Table 8	3.6	4.0	4.2	—
n_1 calculated	5.3	4.8	6.2	5.2
n_1 observed	7	6	9	4
n_2–n_{10} cal. approx.	10.2	9.3	11.9	10.1
n_2–n_{10} observed	11	13	12	13

back to the original size of the quadrats. From the number of species and of individuals there has been calculated the average number of individuals per species and per quadrat; also, on the assumption of the log series, the Index of Diversity (α), and the estimated value of n_1 and the sum of n_2 to n_{10} for comparison with the observed numbers. Unfortunately the information given in this table from the examination of 59 to 75 quadrats does not agree with that on Table 37 for area up to 200 m². The number of species on 60 to 75 m² in the second table is in each case higher than the total for 200 m² given in the earlier table. They must have been taken from different examinations, and it seems likely that greater care was taken in identifying the species on the smaller quadrats. This makes a comparison of the two difficult. For example, the diversity of the population calculated from the quadrats (species and individuals) gives higher values than from the difference in number of species between 1 and 200 m² in Table 37.

In the species with different numbers of individuals it will be seen that the observed value of n_1 is in three cases above the calculated and once below. For n_2 to n_{10} the observed are always above the calculated, but in two of the four cases the difference is less than a single species.

Pidgeon and Ashby, in their original data, distinguish between the perennial and the annual plants. I find that the perennial plants give a closer fit to the log series than either the annuals or the whole flora. The estimated n_1 for the perennials on all plots is 2.7, and the observed 2; the estimated number of all species with up to ten individuals is 7.9, and the observed 8. This may be a coincidence, but it raises the interesting question whether competition within the annuals and within the perennials may not be of a different order from the competition between the two groups.

The number of species in different numbers of quadrats in Raunkaier's system is shown in Table 39 and Fig. 36. On the figure there has also been shown by asterisks the approximate numbers of species in Groups I and V obtained by extrapolation in Figs. 28 and 29. It will be seen that while the relative values are in each case correct, the observed figures for Group I are higher than the estimates, and those for Group V are too small. A departure from the theoretical numbers in this direction would be expected if there was a lack of uniformity in the local environment.

TABLE 39. *Data on plants in New South Wales, classified into Raunkaier's five groups*

	No. of quadrats	No. of species	Raunkaier's groups				
			I	II	III	IV	V
West fenced	50	35	19	6	3	2	5
West unfenced	50	30	19	2	4	2	3
South fenced	40	35	23	6	3	0	3
South unfenced	39	32	17	5	4	1	5

There is no striking difference between the fenced and the unfenced area, but the diversity of the fenced areas is in each case slightly larger than in the corresponding unfenced area.

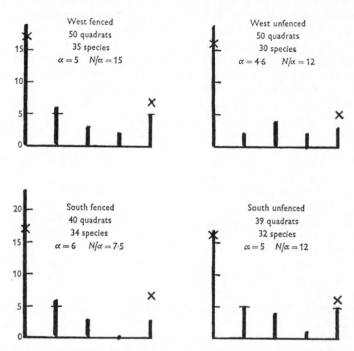

FIG. 36. Pidgeon and Ashby's data for plant quadrats in New South Wales classified in Raunkaier's five groups of abundance. (From *J. Anim. Ecol.* (1950), **38**, 133.)

THE PATTERN OF THE WEED FLORA IN A CEYLON TEA PLANTATION

In 1947, Bond published an account of the distribution of weeds in a tea estate in Ceylon, with special reference to the application of the logarithmic series and the index of diversity. All the data were in reference to quadrats of 50 cm square (0.25 m²). Ten random quadrats were examined in each of twenty-seven plots in each of 2 years. The average number of species of weeds observed in different combinations of quadrats is shown in Table 40.

When the number of species is plotted against log area as in Fig. 37 a nearly straight line is obtained of which the slope corresponds to a diversity (α) of 3.4. Using the formula $S = \alpha(\log_e N - \log_e \alpha)$, derived from the log series, and making $\alpha = 3.4$, Bond shows that there should be approximately ninety-one individual plants, or plant-units, per square metre. If the calculated curve for the log series relation between individuals and species with $\alpha = 3.4$ and ninety-one individuals per square metre is superimposed on the diagrams as A in Fig. 37, we see what a close fit this gives to the observed values. Bond pointed out that, at the time of the quadrat survey, no count was made of the number of individual plants, but a careful re-examination a few months later gave an average of about 566 individual plants per square metre. This gives the relation shown in Fig. 37B. He suggested that the large discrepancy might well be due to the fact that the dominant weed in the area, *Polygonum nepalense* Meisn., according to a "point quadrat" method of

TABLE 40. *Relation between the number of quadrats and number of species of weeds in samples taken in a tea plantation in Ceylon by Bond* (1947)

No. of quadrats	Area in sq m	Average number of species
1	0.25	6.64
10	2.5	13.44
90	22.5	20.83
270	67.5	26.00
Total 2 years		29

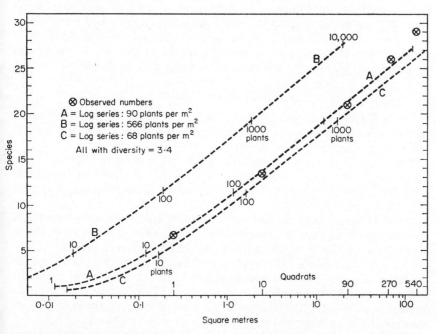

FIG. 37. Increase of number of species of weeds with area sampled in a tea plantation in Ceylon as observed by Bond in 1947, with calculations based on the assumption of a log-series frequency distribution, and with three different levels of average numbers of plants per square metre.

P.B.N.–D

estimation, accounted for about 93 per cent of the total cover, leaving only 7 per cent of the area for all the remaining weed population.

Bond also states that this *Polygonum* is not truly a member of the weed association, as it is a valuable ground cover and has been consistently protected for several years. It is thus a part of the artificial environment of the weeds, just as are the tea trees, and not of the weed association. If therefore the number of species per quadrat is reduced by one (the *Polygonum*) Bond gets an estimate of sixty-eight plants per square metre, the curve for which is shown as c in Fig. 37. It will be seen that this is a very close fit to one species below the original numbers. The number of sixty-eight plants is actually 12 per cent of the total of 566 individuals per square metre, estimated by the "cover" method, and not the 7 per cent, but he pointed out that the *Polygonum* plants are on an average larger than the weeds, and so the allowance necessary for the *Polygonum* might be smaller in individuals than in cover.

Although much of this is theoretical and presumptive, it leads one to consider more closely the difference between plants which are cultivated and the "weeds" which are growing accidentally among them. For example, the weeds in a wheat field (which, by the way, raises new problems as an annual crop) could be considered as an association with the wheat as part of the environment. The weeds in a permanent wheat field (as, for example, the experimental plot on Broadbalk, at Rothamsted Experimental Station) would be more likely to reach an approximate balance than those in a field planted with wheat in rotation. In either case, however, any species of plant of which the seeds were likely to occur as impurities in the wheat seed, would have to be considered separately, much as immigrant species in an animal association.

DESERT FLORA IN ARIZONA

In 1937, Shrieve and Hincley published an account of three botanical surveys, at intervals of a number of years, in the Arizona desert near Tucson in the United States.

The first visit was in 1906, and later counts were made in 1928 and 1936. Owing to the very sparse vegetation it was possible to count the number of individual plants. The results for five areas, each 100 metres square (10,000 m^2), are shown in Table 41.

It will be seen that at the first examination in 1906 there were from six to fourteen species of plant, with an average of 10.2, on each of the areas. These were represented by 27 to 166 individuals with an arithmetic mean of 103, or one plant for each 99 m^2. Twenty-five years later the average number of species had fallen very slightly to 9.6, but the arithmetic mean number of individuals had risen to 147, or one plant to 68 m^2. In 1936 the counts were eight to eighteen species per plot, with an average of 12.2, represented by 193 to 440 individuals, or one plant per 36 m^2.

The table also shows the diversities of population, calculated on the assumption of the log series. Those for the averages of the five plots are calculated from the geometric mean of the number of individuals and not from the arithmetic mean. It will be seen that at the first examination the average

TABLE 41. *Species and individuals of plants in five areas of 10,000 m² each, in arid country near Tucson, Arizona, at intervals of a number of years*

Area		S,11	S,12	S,15	S,16	S,17	Average
1906	Species	12	9	14	6	10	10.2
	Individuals	136	166	97	27	89	103
	Sq. m. per plant	73	60	103	370	112	97
	Diversity	3.0	2.1	4.8	2.8	1.9	2.97
1928	Species	8	13	14	6	7	9.6
	Individuals	116	143	142	69	269	148
	Sq. m. per plant	86	70	70	145	37	68
	Diversity	2.0	3.5	3.8	2.0	1.4	2.35
1936	Species	11	13	18	11	8	12.2
	Individuals	193	235	440	224	302	279
	Sq. m. per plant	52	43	25	45	33	36
	Diversity	2.5	3.0	3.8	2.3	1.5	2.66

diversity was 2.97; at the second visit 22 years later it was 2.35; and at the third, after 8 more years, it was 2.66.

Thus the sequence of development appears to have been that, in spite of a steady increase in the number of plants, from 103 to 279 on the 10,000 m², there is no evidence of an increase in specific diversity. The evidence shows that the number of species has not even increased in proportion to the number of individuals. This might support the idea that the number of individuals within suitable species was not limited by competition; but that the number of possible species was limited by the severity of the physical environment.

Chapter 5

PROBLEMS OF SPECIES AND LARGE AREAS

THE frequency distributions of species with different numbers of individuals, and the rate of increase of number of species with sample size, so far discussed have been based chiefly on samples from comparatively small areas where there is a probability that the whole population belongs to a single association.

When the area of a botanical survey is increased from a few square yards to hundreds of square miles, or a random sample of insects is taken from a number of traps spread over a large area, there is a stage at which the measure ceases to come from a single association, but is drawn from a more and more complex environment. It may be possible—particularly in plants which do not move—to get areas, such as steppe country or semi-desert, with moderately uniform conditions over hundreds of square miles. Normally, however, a single type of environment, producing an association of animals and plants, covers a much smaller area. So, sooner or later, as we increase the area under study we encroach on new environments and associations. This new complexity would be expected to give a greater increase in species number than if the area had been extended within the same environment. An acre of grassland added to another similar acre will produce a much smaller increase in species than if it had been added to an acre of woodland.

The problem of the relation between numbers of species in large areas attracted attention long before the simpler problem of samples within an association had been thought of. As long ago as 1859 Watson gave a table of the areas and the number of species of flowering plants in each of the eighteen provinces and the fifty-eight sub-provinces into which he divided Great Britain, and commented on the fact that 1 square mile of diversified country in the north of the county of Surrey contained nearly half of the number of species found in the whole 760 square miles of the county.

Alfred Russell Wallace, in 1910, gave species numbers for many areas, large and small, in both temperate and tropical climates. Beyond noting the greater rate of increase in the smaller areas, and the greater number of species for similar areas in the tropics than in cooler climates, he was not able to make any new generalizations.

Later Willis (1922) touched on the problem from a slightly different angle, as he was more interested in the frequency of species (and genera) with different areas of distribution, than with the actual number on a particular area. Other approaches overlapping into larger areas have been mentioned in previous chapters.

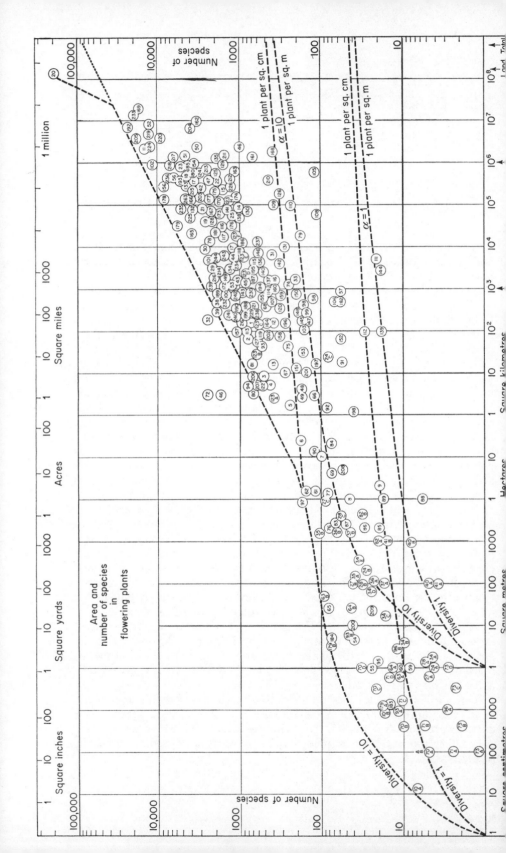

In 1943 I published a diagram showing the relation of species number to area in the flowering plants, over the range of area from 2 in^2 to the whole land surface of the earth, and this has been reproduced in an enlarged and amended form in Fig. 38. It differs in plan from the species-area diagrams already discussed in that both the areas and the numbers of species are on a logarithmic scale, and not only the areas.

The centre vertical line represents 1 km^2 of land. To the right the areas increase geometrically up to the total land area of the earth, which is about 56 million square miles or 145 million km^2. To the left they decrease at the same rate down to 1 cm^2. It may be noted that the total land area is just over 10^{18} cm^2. The numbers of species, on the vertical scale, also increase geometrically from one to a possible limit of 1 million, which is well above any suggested number for the world species of flowering plants.

Each small circle represents a recorded flora placed in the correct position for area and species number. The number within the circle refers to a list in Appendix B (p. 312), which gives the location, area, recorded number of species, and a reference to the source of information. The records range from eight species on 2 in^2 (13 cm^2) up to an estimate of about 200,000 for the flowering plants of the world.

The main error in the recorded numbers of species is one of under-estimation due to insufficient knowledge. This error is small in small areas and in well-studied countries, but may be very large in some tropical area where no extended study has ever been made. There are two possible sources of overestimation: first mis-identification, and secondly the inclusion in some floras of exotic forms which have been brought in only in recent times by agriculture and commerce. It is, however, unlikely that either of these will outweigh the effect of lack of knowledge.

It should be noted in this connection that the logarithmic scale is somewhat insensitive to small changes, particularly in large numbers. For example, in Fig. 38, when the recorded number of species is at the centre of the enclosing ring the upper and lower edges of the ring allow for an increase or decrease of approximately 14 per cent in the original estimate.

The available records are, unfortunately, not quite evenly distributed and there is a particular lack of areas between 2 acres and 1 square mile. Also there are fewer records of the "quadrat" size from the tropics or the subarctic climates. At the other end of the scale I have only been able to find estimates of the total flora for two continents, Europe and North America.

Superimposed on the diagram are dotted lines to show the expected rate of increase, within single associations, calculated from the logarithmic distribution. Examples of this are given for diversities of 1 and 10, and for two

FIG. 38. The number of species of flowering plants recorded in floras of areas of different sizes from all parts of the world. The horizontal ordinate shows the area concerned on a log scale ranging from 1 sq cm on the left to the total area of the world on the right. The vertical scale, also on a log scale, is the number of species recorded in each particular area. (For fuller discussion, see text. For key to localities represented by numbers on the diagram, see Appendix B, p. 312.)

densities of population, one plant per square centimetre and one per square metre. The former is nearer the truth with small plants as in a grassland association, and the latter in woodland or forest. In semi-desert areas there may at times be only one plant per 100 m² or more. It is interesting to note that for this logarithmic distribution with a diversity of 10 and one plant per square centimetre, if the whole surface of the earth were a single plant association, one would expect a total of only about 500 species for the world flora.

Returning to the distribution of the observed points on the diagram, it will be seen that they are chiefly scattered over a narrow belt across the diagram. The richer floras, temperate and tropical are towards the top and the poorer, desert, subarctic and oceanic floras are lower down. Number 111 with only twenty-one recorded species on 2500 square miles is the Island of Kerguelen, subantarctic and far south in the Indian ocean. Number 80 with 700 species in 1.25 square miles is a portion of the Wye Valley in Herefordshire, England, noted among botanists for the exceptional richness of its flora. Numbers 46 and 72 are two areas in Java recorded by Koorders as being exceptionally rich, but from the very brief description of the method of sampling, it would seem possible that the area searched was underestimated. At the lowest end 10A is a record of eight species on 2 in² of grassland in southern England recorded by Blackman.

The upper limit of most of the records has been indicated by a heavy dotted line, and this can be divided into three distinct parts. First, a range A to B (Fig. 41) of rapid proportional rise, slowly falling away from the smallest record to areas of 1 or 2 acres. This changes to a long range of more rapid but steady increase, B to C, from about 10 or 20 acres up to the limits of large continental areas of 5 million square miles. Finally, there is an even more rapid rise, C to D, from the continental areas to that of the whole earth. These three portions appear to be explicable in the following way. As can be seen from the diagram, the first part of the curve A–B, very closely follows the relation required by the logarithmic series with a diversity of 10 and one plant per square centimetre. This portion therefore shows the increase expected with increased sample size in a uniform population, or a single ecological association. On an average the physical environment does not change much as the area sampled is increased within these limits. In this range, as already discussed, in all except very small samples the multiplication of the area by a constant factor adds a constant number of species. The relation is given by

$$S = \log_e \alpha \left(1 + \frac{N}{\alpha}\right).$$

As soon as we pass much beyond this limit we become more and more likely to include new ecological conditions within the area, with the result that the number of species will increase more rapidly than would be expected if the environment sampled remained uniform. This is the condition in the range B–C of the limiting curve where there appears to be a straight-line relation between log area and log species. The formula suggested for this is

$$\log S = k \log N + \log p$$

where k is a measure of the slope of the line, or of the rate of increase of log species with log area, and p is the point at which the line crosses the unit measurement of area (see Fig. 40). As this relation does not hold for areas much smaller than 1 km² it is convenient to take this as a unit, in which case p would be the expected number of species on 1 km².

It will be seen that, on this assumption, for the limiting line in the diagram, p has a value of about 450 species and k about 0.26. From this it follows that an increase of the log area by 1 (i.e. a tenfold increase in area) produces an increase in log species of 0.26, which is a multiplication of about 1.8. To make these figures comparable with previous ones, a doubling of the area increases the number of species by about 20 per cent.

When we add continents together to get a flora of the whole world an entirely new factor comes in, that of the evolutionary history of the floras. Fewer species are common to two continents than to two halves of the same

TABLE 42. *Number of species of flowering plants in relation to area in France (Cailleux)*

	Area km²	Species
Whole of France	550,000	4400
Paris region	30,000	2871
Pas de Calais	6605	1400
Arondissement Hazebrouck	708	1010
Environs of Strasbourg	140	980
Le Fazel, Oise	7	516

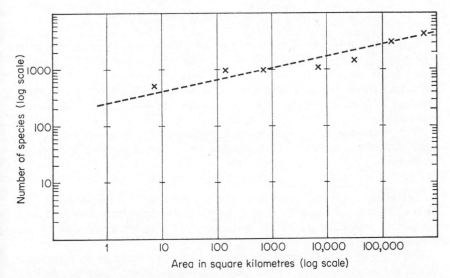

FIG. 39. Relation between area and number of species of flowering plants, each on a log scale, in France from data from Cailleux.

continent, as the floras have evolved differently. Therefore on adding two isolated continental areas together we get a sudden increase far beyond that due to the increased ecological variety which is introduced. This is the last steep portion of the curve, c to d.

Thus the inverted S-shaped form of the limiting line appears to be consistent with the known facts of sampling effect, ecology and evolution.

At a colloquium in Paris in 1952, Cailleux (in a typewritten sheet) drew attention to the approximately straight-line relation between log area and log number of species in a comparison of the floras over a large range of area in France. His data are shown in Table 42 and Fig. 39. At this time, and also in Cailleux (1953), he proposed for this relation a formula which is identical with that given above except for different nomenclature of the constants. Using my terminology, and assuming the significance of the straight line, we find that an increase of 6.0 in the log area is associated with an increase of 1.26 in the log species. Hence $k = 0.21$ and $p = 250$ species, as compared with 0.26 and 450 for the limiting line in my world diagram. The slopes of the lines, measuring the note of increase, are very similar: the higher value of p in my diagram is to be expected, as the line is drawn at the upper limit of richness of flora.

The changing relation between number of species and area of the sample size in plant populations, as the area is increased from a few square centimetres up to the total land surface of the globe, can therefore be summarized as follows:

I. In very small samples of a few hundred individuals the relation is best expressed by the logarithmic series which gives

$$S = \alpha \log_e \left(1 + \frac{N}{\alpha}\right) \text{ or } 2.30 \, \alpha \log_{10} \left(1 + \frac{N}{\alpha}\right).$$

This gives, throughout the range in which the unit 1 is significant in relation to N/α, a curved line (Fig. 41A) for the relation between S and log N, which tends to straighten out sooner or later according to the number of individual plants per unit area and the diversity of the population. Greater diversity or fewer plants per area both increase the area at which the change takes place.

II. For samples larger than those considered in I, but taken within the same association, the unit can be neglected in relation to N/α and the formula becomes

$$S = \alpha \log_e (N/\alpha) \text{ or } 2.30 \, \alpha \log_{10} (N/\alpha).$$

This gives a straight-line relation between S and log N, the slope depending on the diversity, and implies that, throughout the range of this straight line, a constant *multiplication* of the area produces a constant *addition* to the number of species.

If the log series is accepted this should continue as long as the increasing size of the sample still lies within the original association. If the distribution were a log normal, the curve would be slightly sigmoid, with a flattening out at higher levels where the rate of increase of species would become smaller.

If the data for this range are plotted on a log S × log N diagram the straight line becomes a curve, as shown in Fig. 41B, which very slowly flattens out, but never becomes a true straight line.

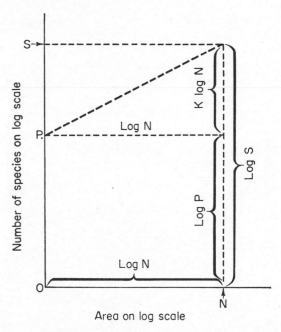

FIG. 40. Relation between area and number of species for larger areas where both area and number of species are on a log scale.

III. When the increased sample size reaches the limit of the single environment or association (which may be after a few square yards or some hundreds of square miles according to the diversification of the physical environment) the rate of increase of species will accelerate, and the observed situation is shown in Fig. 40 plotted on a log × log scale. It is a straight line for which the formula can be expressed as

$$\log S = k \log (N/x)$$

where k is a measure of the slope of the line (corresponding, but not equivalent, to the diversity α in the previous figure) and $\log x$ the point at which the extrapolation of the line downward to small area would cross the ordinate $\log S = 0$, or $S = 1$. No theory has yet been developed to account for this relation, which is well shown in Fig. 41D. This relation apparently extends to the limits of large areas, such as continents, within which there has been some uniformity of origin of the flora, and the possibility of interchange and spread within the areas.

IV. When we pass this limit, by adding continents together to produce a world flora, we are adding floras not only with more and more diverse physical environments, but also with differing evolutionary development through comparative isolation over geological ages. A square mile of natural forest land in Europe added to a similar area even in a distant part of the same continent will produce a smaller total number of species than if it were added to a square mile of forest in North America.

Thus the final increase from continental to world flora (the latter covering about 56 million square miles) gives a still greater rate of increase as shown in Fig. 41E. This is undoubtedly the effect of extreme diversity of origin due to long isolation.

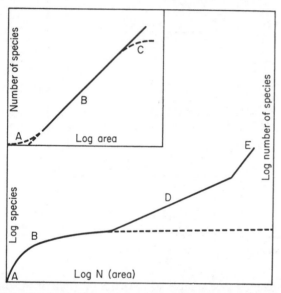

Fig. 41. Relation between log area and the number of species of plants (the latter on an arithmetic and a log scale) to show the difference between smaller areas where the logarithmic series appears to hold and the large areas where the rate of increase of species is more rapid. The lower part of this diagram should be compared with Fig. 38.

THE PATTERN OF THE WORLD POPULATION OF INSECTS

Since so many random samples of different groups of animals, taken by different methods and in different parts of the world, have indicated regular mathematical patterns in the relative abundance of species, it is perhaps interesting to speculate as to what might be the pattern of all the insect species in the world at any one time, and what might be inferred from it about the absolute abundance of species.

For such an investigation it is necessary to have estimates, based on some evidence, for two or three constants, of which perhaps the two most usually available are the total numbers of individuals and of species. If we make the provisional assumption that the frequency distribution of abundance is a logarithmic series, then these two constants are sufficient to define the series and all its properties. For any one number of species and one number of individuals only one logarithmic series is possible. If we postulate a log-normal distribution, a third constant or parameter is needed, which may well be the measure of the spread of the distribution round the mean, as given by the standard deviation.

The total number of species of insects at present known is larger than that for any other group of animals. In fact, in the British Isles the number of known species, over 20,000, is greater than the total in all other groups of animal and plants together. Various estimates of the number of species already described in the whole world usually suggest about 1 million, but Metcalf (1940) suggests over 1.5 million. New species are being described at the rate of about 10,000 per year. But in view of the numerical preponderance of rare species in all populations, and of the many groups of insects that have scarcely been studied, we must assume that, for the world as a whole, most of the really rare species are still unknown. My own opinion is that certainly less than half of the existing species of insects have yet been described, and that the proportion may be well below this. I suggest that we might take 2 million as a low estimate, and 3 million as quite possibly closer to the truth.

To try to estimate the number of individual insects existing at any one moment is an even more questionable occupation. Reliable evidence, or any applicable over large areas is difficult to find. In a series of careful extractions of insects from the soil at Rothamsted Experimental Station, in S.E. England, the numbers rose, with improvements in technique, from an early estimate of 20 million to later figures of 200 to 400 million per acre; which is between five and ten insects per square centimetre of the surface. A recent estimate (Morgan in lit.) of the number of Chironomids emerging from 1 yd^2 of the mud at the bottom of a shallow loch in Scotland, was up to 20,000–30,000, or two to three per square centimetre. We do not yet have sufficient information about the density of population in tropical countries, nor yet in the hot and cold deserts. Nor have we any accurate measurements of the total number of insects above the ground in a forest community. Many places must be more prolific than S.E. England, and many less so. Since the total land surface of the earth is about 50 million square miles, the number of square centimetres is of the order of 1.3×10^{18}. From our British estimate this would suggest a world estimate of 8×10^{18}. In view, however, of the great expanse of cold areas in the arctic and subarctic lands, and of hot deserts in the subtropical parts of the world, this is probably an overestimate. It would seem likely, however, that the total is not far above 10^{18}, or 1 million million million, so we will take this as a working hypothesis.

From 10^{18} individuals and 2 or 3 million species we get arithmetic means of one-half or one-third of a million million individuals. It must be remembered, however, that as the distribution is almost certainly on a geometric scale of abundance this does not mean that half the existing species have more, and half less, than this number of individuals. The median is not at the arithmetic mean, but probably near the geometric mean, and so very much lower. The question of the most abundant species will be discussed later.

Is this estimate for the arithmetic mean absurd? I have myself seen a single swarm of locusts, for which a conservative estimate gave about 10,000 million (10^{10}) individuals, and this was not an outstandingly large swarm. During a migratory flight of butterflies (*Pyrameis cardui L*) in California McGregor (1924) estimated that there were about 3000 million individuals concerned. In 1939, Oliver (1943) recorded a vast number of ladybirds (*Coccinella undecim-punctata*) coming inshore from the Mediterranean on the Egyptian coast west of Alexandria. Those that had fallen into the sea and were washed ashore formed a drift line at least 14 miles long, with about 70,000 beetles per foot. This gives a figure of about 4500 million just for those lost from the swarm, and only on the observed stretch of shore.

But all these insects are comparatively large; it is among the small species, such as Collembola, and among the aquatic insects such as may-flies and midges, or among the ants and termites, that the large populations will be found. It would seem likely that the number of mosquitoes or related swamp insects in some subarctic areas must run into many millions of millions. At one insect per square centimetre it requires only a single square mile to produce 26,000 million individuals. Large populations are also likely to be found in small species with wide distributions.

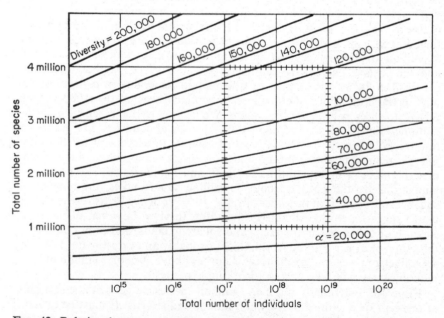

FIG. 42. Relation between number of individuals and number of species in an animal population, on the assumption of the logarithmic series, where both number of individuals and number of species are very large.

WORLD INSECTS ACCORDING TO THE LOGARITHMIC SERIES

To illustrate the conditions in a world population of insects if it were distributed in a logarithmic series, Fig. 42 shows the relation between very high values of N and S and the index of diversity, which in this case is practically identical with n_1, the number of species with one individual. The area likely to be of interest in the present problem is indicated in an inner rectangle.

It will be seen that with 10^{18} individuals and 2 million species there should be a diversity—and so a number of species with one individual—of 65,000. This is presumably those that are either just originating, or just dying out. If we take 3 million species, the number rises to 100,000, in each case the proportion being about 3.3 per cent of the total.

When the average number of individuals per species is so large the value of x is extremely near to unity; with $N = 10^{18}$ and $S = 3$ million, $x = 0.999,999,999,999,899,8$! Thus it is possible to assume that at least the first half of the series will be indistinguishable from a hyperbolic series. Using the approximation that the sum of a hyperbolic series up to the rth term is

$$n_1 (\log_e r + 0.5772)$$

we get the number of species with different numbers of individuals, summed up to different powers of 10, as shown in Table 43. It will be seen that the median, with 50 per cent of the species above and 50 per cent below, is at 2,187,000 for 2 million species and at 1,778,000 for 3 million. It is interesting to compare these with the arithmetic means of 5×10^{11} and 3.3×10^{11} respectively.

Some years ago the late Dr P. Grundy gave me a formula (reproduced in Gower, 1961) for the median of a logarithmic series as

$$\frac{0.56146}{\sqrt{1 - x}} + 0.81524.$$

TABLE 43. *Frequency distribution of species with different numbers of individuals in a population of animals with 10^{18} individuals, and 2 or 3 million species, based on the log series*

Species with up to this number of individuals	Total species out of			
	2 million		3 million	
	Number	Percentage	Number	Percentage
1	66,000	3.3	100,000	3.3
10	193,000	9.7	294,000	9.8
100	342,000	17.1	520,000	17.3
1000	493,000	24.7	750,000	25.0
10,000	645,000	32.2	981,000	32.6
100,000	797,000	39.8	1,211,000	40.4
1,000,000	949,000	47.4	1,442,000	48.1
1,778,000	—	—	1,500,000	Median
2,187,000	1,000,000	Median	—	—
10 million	1,100,000	55.0	1,673,000	55.7

This gives a median of 2,185,000 for 2 million species and 1,773,000 for 3 million, and so does not differ significantly from my calculation on the assumption of the hyperbola. The percentage of species up to each level is almost independent of the total number, with only a very slight increase from 2 to 3 million.

Gower (1961) has discussed the estimation of the total number of species up to very high values for individuals per species, where it would be wrong to assume the similarity with the hyperbolic series, and has produced some general formulae which he has applied to the case of 10^{18} individuals in 3 million species. His values, up to 10^7 individuals per species differ from my estimates by not more than 0.1 per cent. For the higher values his results are summarized, as percentages, in Table 44. They suggest that less than 0.05 per cent of the species have more than 10^{14} (100 million million) individuals. This percentage gives about 1500 out of a total of 3 million, or 1000 out of a total of 2 million species.

TABLE 44. *Estimate of percentage of total species with up to certain high levels of individuals per species, in populations based on the logarithmic series*

Up to this number of individuals per species	Percentage of total species
10^7	55.8
10^8	63.5
10^9	71.2
10^{10}	78.9
10^{11}	86.5
10^{12}	93.9
10^{13}	99.3
10^{14}	99.95

The estimate of 3.3 per cent of the total species in the world being at any moment either just emerging or just dying, and 10 per cent with not more than ten individuals, seems to be very high when it is considered as a proportion of the species. When, however, the proportion of individuals is considered, the number is almost infinitesimal. In the case of 2 million species, out of the assumed 10^{18} individuals only 66,000 will belong to species with a single individual, and only ten times this number to species with ten individuals or less. Thus one would have to take a random sample of 15 million million insects to have an even chance of a single incipient species, or 1.5 million million before getting one of a species with ten or less.

I know of no work that has produced any direct or indirect evidence on the number or proportion of incipient or dying species in the world, but in conversation Professor J. B. S. Haldane has said that he might have expected a still higher number.

It is interesting to note that, on the assumption of the close approximation of the log series to the hyperbolic series in these extreme cases, the number of individuals in species up to the median abundance would, for 2 million species, be 2,187,000 × 66,000 or 1.44 × 10^{11}. Thus the number of individuals contained in the 50 per cent least-common species would be only about 1.4 in every 10 million of the population. The true log series could only differ from the hyperbolic series in having a *smaller* number of individuals.

WORLD INSECTS ACCORDING TO THE LOG-NORMAL DISTRIBUTION

The log-normal distribution (see p. 16) can be defined by three constants or parameters which, for our present purposes, are most conveniently

 (1) N = the total number of individuals,
 (2) S = the total number of species,
 and (3) σ = the standard deviation expressed as a power of "e".

The interrelation of these three with the median on the log scale (a) is given by the fact that the arithmetic mean number of individuals per species

$$N/S = a\, e^{\frac{1}{2}\sigma^2}.$$

If we make, therefore, any assumptions about the number of individuals and of species, we get, for each pair of such values, a number of possible alternative pairs of values for σ and a. For example, if we take $N = 10^{18}$ and $S = 2 \times 10^6$ then the arithmetic mean number of individuals per species is 5×10^{11}, or half a million million. If we suggest that the S.D. might be e^3, it follows that $e^{\frac{1}{2}\sigma^2} = 90$, and so the median of the log-normal distribution of the 2 million species would be at 5.5×10^9 or $e^{22.4}$. Otherwise half the species would have more, and half less, than this number of individuals.

Table 45 shows calculated values for the medians for 10^{18} individuals when the number of species is taken as either 2 or 3 million, and the S.D. as a power of e is given the values 2, 3, 4 or 5. Figure 43 shows the same data set out graphically and also includes the number of species in each S.D. range from + 5 S.D. above the median to — 5 S.D. below, and the small numbers left outside this range. The total population of 10^{18} individuals is indicated by the upper broken line, and the lower limit of one individual per species by the lower broken line. Some of the theoretical groups extend beyond these limits, because the log-normal distribution allows of fractional species: in practice such distribution would be truncate. There is little difference in the position of the median when the number of species is increased from 2 to 3 million, but the number in each S.D. class is increased by 50 per cent.

Taking the lower value of the S.D. at e^2 it would appear that, with the median at 6.75×10^{10} individuals per species, less than one out of the 2 million species would be represented by fewer than 3 million insects. This seems to be much too high a lower limit, and makes no allowance for very rare, very local, or incipient species.

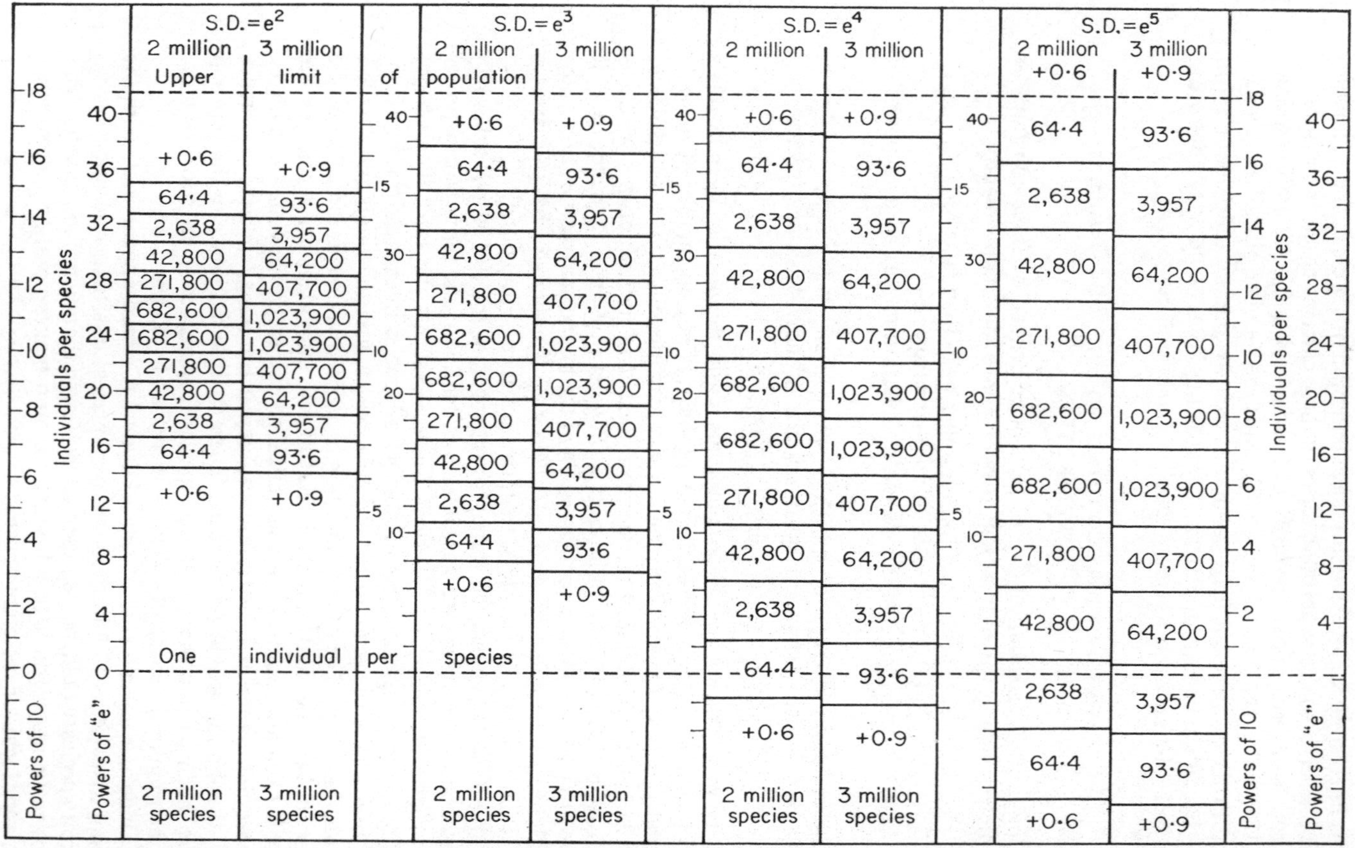

S.D.=e^2 Upper limit		S.D.=e^3		S.D.=e^4		S.D.=e^5	
2 million	3 million	2 million	3 million	2 million	3 million	2 million	3 million
+0·6	+0·9	+0·6	+0·9	+0·6	+0·9	+0·6	+0·9
64·4	93·6	64·4	93·6	64·4	93·6	64·4	93·6
2,638	3,957	2,638	3,957	2,638	3,957	2,638	3,957
42,800	64,200	42,800	64,200	42,800	64,200	42,800	64,200
271,800	407,700	271,800	407,700	271,800	407,700	271,800	407,700
682,600	1,023,900	682,600	1,023,900	682,600	1,023,900	682,600	1,023,900
682,600	1,023,900	682,600	1,023,900	682,600	1,023,900	682,600	1,023,900
271,800	407,700	271,800	407,700	271,800	407,700	271,800	407,700
42,800	64,200	42,800	64,200	42,800	64,200	42,800	64,200
2,638	3,957	2,638	3,957	2,638	3,957	2,638	3,957
64·4	93·6	64·4	93·6	64·4	93·6	64·4	93·6
+0·6	+0·9	+0·6	+0·9	+0·6	+0·9	+0·6	+0·9
One individual		per		individual			
2 million species	3 million species	2 million species	3 million species	2 million species	3 million species	2 million species	3 million species

Powers of "e": 40, 36, 32, 28, 24, 20, 16, 12, 8, 4, 0 — Individuals per species

Powers of 10: 18, 16, 14, 12, 10, 8, 6, 4, 2, 0

FIG. 43. Theoretical frequency distribution of species with different numbers of individuals per species, in populations following a log-normal distribution, with total numbers of individuals = 10^{18}; with total species 2 or 3 million; and with standard deviations e^2, e^3, e^4, and e^5. These being considered as a possible basis for the insect population of the whole world.

With S.D. $= e^3$ there would be less than one species in the world with under 15,000 individuals and if the number of species is increased to 3 million there is little alteration in these limits.

If we take the higher estimate of e^5 for the S.D. we get at the upper limit about one species with more individuals than the total population, and at the other end about 2500 species with less than one individual. These are below what Preston (1948) called the veil line and suggest a truncate distribution.

On the whole, if the log-normal distribution is to apply, it seems that an estimated S.D. of about e^4 is reasonable. This gives for the 2 million species, about sixty-five with twelve individuals or less, and at the upper end, perhaps one species with over 10^{17} individuals.

Some slight confirmation of this approximation may be obtained from known large samples or censuses of animal populations. In the insects we have no samples reaching even 1 million, but 32,000 Lepidoptera in a light trap in S.E. England including 285 species (Williams, 1953a) gave, when fitted to a log normal a S.D. of $e^{2 \cdot 2}$. In the birds, the distribution of about 63 million British land-nesting birds (p. 45) with 142 species gave a S.D. of $e^{3 \cdot 2}$. An estimate of the whole nearctic bird population by Preston (1948) with 10^{10} individuals in 606 species gave $e^{3 \cdot 7}$. Another calculation by Preston (1958) of 25 million birds in an Audubon Christmas Survey with 487 species, gave $e^{3 \cdot 5}$; and a total of 83.4 million birds in 7 years of the Audubon censuses with 600 species gave $e^{4 \cdot 1}$. It seems likely that the standard deviation will increase with the greater area over which the population is spread, as some species will be universal and others very local, thus remaining in small numbers even when a very large total area is included.

On the basis of S.D. $= e^4$, Fig. 44 shows the pattern of abundance of the world insects. The background curve and histogram is a log normal showing the limits of the successive departures above and below the median of 1 S.D. (e^4), and the percentages and the actual number of species out of the 2 million in each of these classes. About two-thirds of the total species will lie between the limits of $+ 1$ S.D. and $- 1$ S.D. or between 3 million and 9000 million individuals per species.

One of the properties of the log-normal distribution (Aitchison and Brown, 1927, p. 12) is that if it has a S.D. of e^σ and a median at a, then the first moments have a distribution in a log-normal with the same S.D., but with the median at $a \times \sigma^2$ on the log scale. It is important to note that the first moments are the number of units or individuals, and their distribution is the number of individuals in all species in the same abundance class. Thus it is possible from the distribution of the species, if in a log normal, to infer the distribution of the individuals associated with them at any level of abundance.

As applied to the present problem, a population of 10^{18} individuals in 2 million species with a S.D. of e^4 has its species median at $e^{18 \cdot 9}$ (Table 45). It follows that the median of the distribution of individuals will be at $e^{18 \cdot 8 + 16} = e^{34 \cdot 9}$. The S.D. of this is still e^4. Figure 45 shows the distribution of species and individuals in this population divided into corresponding groups so that

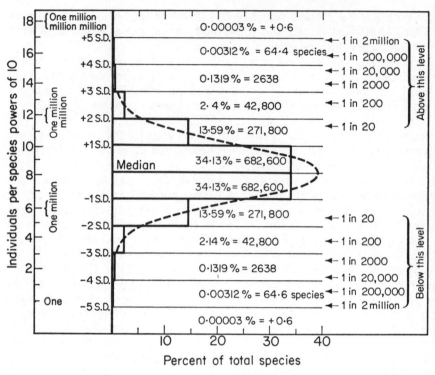

Fig. 44. Frequency distribution of an insect population on the basis of a log-normal distribution, with 10^{18} individuals, 2 million species and a standard deviation of e^4; showing also the number and proportion of the species at different levels of abundance.

TABLE 45. *Value of the median abundance for populations with 10^{18} individuals and 2 or 3 million species, with different values of the S.D., in populations based on a log-normal distribution*

Values of S	2 million	3 million
Values of S/N	5×10	3.3×10^{11}
Values of $\sigma \quad e^{\frac{1}{2}\sigma^2}$	Values of median for above values of N and S	
2 7.4	$6.75 \times 10^{10} = e^{24 \cdot 9}$	$4.46 \times 10^{10} = e^{24 \cdot 5}$
3 90	$5.5 \times 10^{9} = e^{22 \cdot 4}$	$3.76 \times 10^{9} = e^{22 \cdot 0}$
4 2981	$1.68 \times 10^{8} = e^{18 \cdot 9}$	$1.17 \times 10^{8} = e^{18 \cdot 7}$
5 268,000	$1.88 \times 10^{7} = e^{16 \cdot 7}$	$1.23 \times 10^{7} = e^{16 \cdot 3}$

For 10^{17} or 10^{19} individuals the above values for the medians must be divided or multiplied by 10.

at any level of species abundance the number of individuals at that level can be read off.

The following points can be noted. First that the 1 million species below median abundance contribute only 0.00316 per cent of the individuals, although this extremely small proportion actually includes 31.5 million million individuals. There is, on an average, only 0.6 of a species above the level of the $+$ 5 S.D. in the species distribution. Above the corresponding level in the individual distribution there are 15.86 per cent of the total individuals. This problem of the most abundant species is considered in more detail in the following paragraphs.

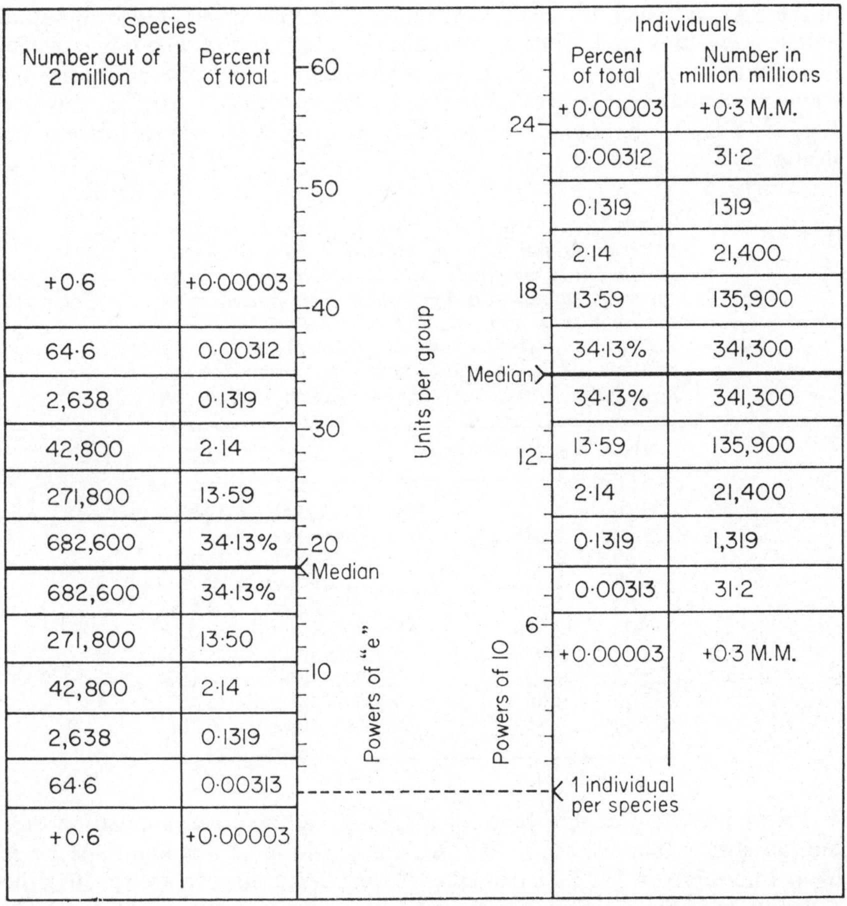

Species					Individuals	
Number out of 2 million	Percent of total				Percent of total	Number in million millions
		60				
				24	+0·00003	+0·3 M.M.
					0·00312	31·2
		50			0·1319	1319
					2·14	21,400
+0·6	+0·00003	40		18	13·59	135,900
64·6	0·00312		Median⟩		34·13%	341,300
2,638	0·1319				34·13%	341,300
42,800	2·14	30		12	13·59	135,900
271,800	13·59				2·14	21,400
682,600	34·13%	20			0·1319	1,319
682,600	34·13%	⟨Median			0·00313	31·2
271,800	13·50			6	+0·00003	+0·3 M.M.
42,800	2·14	10				
2,638	0·1319					
64·6	0·00313				⟨ 1 individual per species	
+0·6	+0·00003					

(Units per group axis in center; "Powers of "e"" and "Powers of 10" labels along vertical axes)

FIG. 45. Numbers and percentages of individuals and of species at different levels of abundance in insect populations distributed log-normally with 10^{18} individuals, 2 million species, and a standard deviation of e^4.

The Most Abundant Species in A Log-Normal Distribution of World Insects

The departures, in terms of the standard deviation, at which different proportions of the total groups (species) are found are given in most mathematical tables in the form of probabilities. In these, the value given is normally that for which a single unit shall be either above or below this departure from the mean. In our present problem we are only interested in the chances of it being above the upper limit, and so the chances are half those given in the tables.

To facilitate further studies I have made Table 46, which gives, for different sized log-normal populations, the departures from the median, in terms of the S.D., at which the chances are that one single species out of the total will be above this limit. Thus in a population of 100 species one single species is likely to be beyond 2.326 S.D. above the median. In 1 million species the single most abundant species is likely to be above 4.76 S.D. from the median. Figure 46 shows the same in graphical form from which approximate intermediate values can be found.

TABLE 46. *Showing the departures above the median, in terms of the S.D., of the single most abundant species in populations with different numbers of species arranged in a log-normal distribution. P = the number of groups in the populations. S.D. = the lower limit of departure from the median in multiples of the Standard Deviation above which the probability is that only one single group will occur*

P	S.D.	P	S.D.
10	1.282	20,000	3.891
20	1.645	100,000	4.27
50	2.054	200,000	4.417
100	2.326	1 million	4.76
200	2.576	2 million	4.892
500	2.878	20 million	5.327
1,000	3.090	200 million	5.731
2,000	3.291	2,000 million	6.109
5,000	3.55	20,000 million	6.467
10,000	3.719	200,000 million	6.806

In the present problem we are dealing with an assumed population of 2 million species, for which we see that the single most abundant species is likely to be above 4.892 S.D. from the median. Since, however (see p. 107), the distribution of the individuals in their log-normal distribution is 4 S.D. above that of the species, it follows that the individuals in this most abundant species will consist of all those beyond 0.892 S.D. above their median. From the published probability tables for the normal distribution this value is one-half of 0.3725 or 18.6 per cent of the total number of individuals. So in the assumed population of 2 million species with 10^{18} individuals we might expect the single most abundant to have about 186,000 million million individuals.

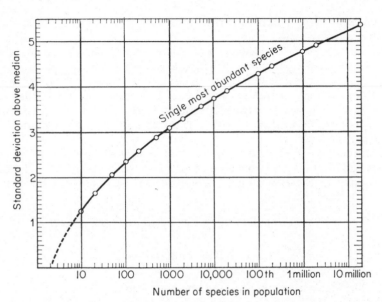

Number of species in population

FIG. 46. Diagram showing the relation between the number of species in a population and the distance above the mean (measured in terms of the standard deviation) of the single most abundant species.

THE MEASUREMENT OF RELATIONSHIP BETWEEN DIFFERENT POPULATIONS WHEN ONLY THEIR RELATIVE SIZES BUT NOT THE NUMBER OF INDIVIDUALS IS KNOWN

If we assume the existence of a log-series distribution in a population with a diversity of α, then the number of species (S) in a sample containing N individuals is given by

$$S = \alpha \log_e (1 + N/\alpha)$$

If N/α is large, as it should be for a reliable sample, then it is possible to neglect the "1" and we get

$$S = \alpha \log_e N/\alpha$$

This relation holds for that portion of the curve of relation between S and log N which is a straight line, but not for very small samples and possibly not for very large samples from the same association. It is, however, important to note that this straight-line relation has been demonstrated to exist over a considerable range of sample size without the assumption of the log series or any other distribution.

Assume that we wish to compare two samples, with A and B individuals and a and b species respectively; then, if they were random samples from the same population, not only would they each have the same diversity, but when added together in a single sample the diversity would still be the same. So the problem of ascertaining similarity of origin is simple if we know the number of individuals.

If, however, we are dealing with plant populations, it is not always possible to count the number of individuals, as has already been discussed; nor is it always possible in large populations of animals. If, however, we can make the assumption that the number of individuals present (whatever its actual level) is on an average proportional to the size of the area sampled, then it is possible to deal with the relationship of plant population by the following method.

Consider two areas of size A and B containing respectively a and b species. If they had been samples from a single population the total number of species, T, expected in the two samples combined can be obtained from the two following relations:

(1) The increase in number of species by adding area A to B

$$= T - a = \alpha \log_e \frac{A + B}{A}.*$$

(2) The increase in species by adding area B to A

$$= T - b = \alpha \log_e \frac{A + B}{B}.$$

From these two equations T and α can be calculated without knowing the number of individuals per unit area.

The number of species common to the two areas, on this assumption of identity of origin, should therefore be $a + b - T$, and if we know the observed number common to the two (which must follow if we know the identity of the a and b species in the two areas) we can compare this with the calculated estimate.

If they are nearly the same, there is strong evidence that the two areas are closely related in origin, being basically chance samples from a single larger flora.

If, as is more frequent, the observed number is smaller than that calculated, the difference between the two—best expressed as the percentage ratio between observed and calculated—can be a useful measure of relationship. Of course, if there is no relation between the two areas there will be no species in common.

The case where the observed number of species in common is greater than

---

* Since it is often inconvenient in calculation to use natural logarithms, the formula can be written $k \log_{10} \dfrac{A + B}{A}$ in which case $k = 2.30\alpha$ (see p. 67).

the estimated can occur under somewhat special conditions which will be discussed later (p. 117).

The whole argument can be expressed graphically, as in Fig. 47, by using the straight-line relation between the number of species and log area as the base line. If the two samples are random ones from the same population they will lie on the same line of diversity as shown by x and y (representing $a/\log A$ and $b/\log B$) in the figure. In this case T in relation to $\log (A + B)$ is shown by z_1 on the continuation of the same line. If the two floras are not of common origin z will be above the line, at say z_2, when there are too few species in common (and hence too many in the total); or below the line at z_3 when there are too many species in common and hence too small a total.

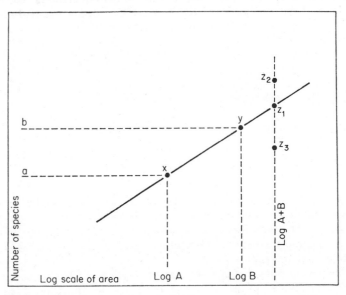

FIG. 47. Diagram to illustrate the relation between the area and the number of species in two localities, and the number of species when the two are combined into one sample.

The relation between sample size and number of species, as so far discussed, applies only when the sample areas are selected each within a single type of environment or plant association. When we increase the size of the area so that it includes more and more different associations (such as is the case if we take an area of several square miles) the rate of increase of species with sample size becomes more rapid. This problem has already been discussed (p. 96) and it will be seen from Figs. 38 and 41 that with large areas, from several square miles up to the limits of the continents, the relation changes from $S = k \log A$ to something closely resembling $\log S = K \log A$. The k in this case is not simply related to the diversity, and has at present only an empirical value.

If this relation can be taken as a working hypothesis, these large areas can be compared with each other by changing the two simultaneous equations on p. 112 to

$$(1)\ \log T - \log a = k \log \frac{A + B}{A};$$

$$(2)\ \log T - \log b = k \log \frac{A + B}{B}.$$

In 1944 and again in 1947 I made some calculations on the relationships of some island floras, but unfortunately used the formula suitable only for smaller areas. Here I give, from the same data, calculations made on the assumption of the log species × log area relation. It is curious that in most cases the effect of this change on the interpretation of the results is very small.

The first example studied was the flowering plants of the two islands, Guernsey and Alderney, in the Channel Islands off the north coast of France. Guernsey has 804 species on an area of 24 square miles. Alderney has 519 species on 3 square miles. On the assumption of identity of origin, the two following relations should hold,

$$(1)\ \log T - \log 804 = k \log \frac{24 + 3}{24} = 0.0511\ k;$$

$$(2)\ \log T - \log 519 = k \log \frac{24 + 3}{3} = 0.954\ k.$$

From which it follows that $T = 813$ and hence the expected number of species common to both islands should be 510. The observed number (Marquand, 1901) was found to be 480. This is 94 per cent of the number expected on complete identity of origin, and is in full agreement with all other evidence of a close relationship.

In 1944, Exell published an account of the flora (vascular plants) of the island of San Thomé off the west coast of Africa, to which he added a comparison with two other islands, Principe and Annobon, in the same area, and a brief comparison with the flora of Fernando Po.

The four islands are roughly in a straight line running in a S.S.W. direction from the coast near Mount Cameroon. Fernando Po is about 32 km from the mainland; Principe is about 210 km from Fernando Po; San Thomé is 135 km from Principe, and Annobon is about 180 km farther out. Annobon is about 340 km (210 miles) from the nearest coastline at Cape Lopez, Gaboon. San Thomé is almost exactly on the equator.

The areas of the islands, the number of species recorded for each, and the observed numbers of species common to different pairs of islands are shown in Tables 47 and 48. Figure 48 shows for each pair of islands the relation (as a percentage) between the observed number of species common and that calculated on the assumption of identity of origin. All these are calculated from the relation $\log S = k \log A$, and not from the relation $S = k \log A$ which I used in a previous analysis (Williams, 1947a). As in the case of the two Channel

Islands, the new calculated percentages are very similar to the previous ones and require no alteration in the interpretation of the relation of the island floras.

Taking first the total number of all species (Fig. 48A), we see that the closest relations are between San Thomé and Principe (72 per cent), and between San Thomé and Annobon (71 per cent). Fernando Po is more closely related to San Thomé (50 per cent) than to Principe (45 per cent) although the latter is less far away. The lowest relation (42 per cent) is found between Fernando Po and Annobon, which are the two most widely separated islands.

TABLE 47. *Area and number of flowering species of plants in four islands off the west coast of Africa*

			Species		
	Area sq. km.	Total	excluding endemics	excluding world weeds	excluding endem. and world weeds
Fernando Po	2000	826	727	—	—
San Thomé	1000	556	448	454	346
Principe	126	276	241	223	188
Annobon	16	115	98	83	66

TABLE 48. *Number of species common to pairs of islands as discussed in Table 47. The open numbers include the world-wide weeds, the numbers in brackets exclude these weeds. The inclusion or exclusion of endemics does not affect these figures*

	Fernando Po	San Thomé	Principe
San Thomé	187	—	—
Principe	128	183 (137)	—
Annobon	47	80 (55)	52 (35)

At the time of my earlier analysis Exell suggested that it would be interesting to remove from the floras certain elements which might be interfering with the interpretation of the natural relations. These were firstly a number of widespread tropical weeds whose presence was due largely to commerce, and the actual recording of which on any island was largely a matter of chance. Secondly, there were a small number of species endemic to each island, which must be supposed to have originated there, and not to have spread in from any wider population. The number of weeds for Fernando Po was not available.

It is, of course, obvious that the removal of the endemic species will reduce the total number of species for each island without affecting the number common to any pair. The removal of the widespread weeds, on the contrary, reduces both the total number of species for each island and, to an even greater proportion, the number of species common.

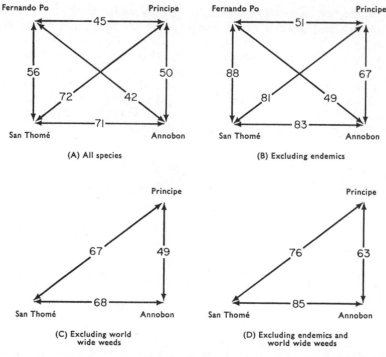

FIG. 48. The relationships between the floras of four islands off the west coast of Africa, calculated from the observed and calculated numbers of species common to pairs of islands.

Figure 48B shows the result of removing all the endemic species. The greatest change is an increase in the relation between the two larger islands, Fernando Po and San Thomé, from 56 per cent to 88 per cent. San Thomé also shows high relationship to both the other small islands. The relation between Fernando Po, Principe and Annobon remains in the same sequence, but all slightly increased in value.

When only the world-wide weeds are excluded (Fig. 48C) there is a very slight reduction in the relationships in the three islands for which data are available, and there is no significant difference in interpretation.

When both world-wide weeds and endemics are removed (Fig. 48D) there is a small increase in all the relations, but the sequence of resemblance remains the same.

To summarize the results, taking first the three outer islands, we infer from all four calculations, with or without endemics and weeds, that the relation of San Thomé to Principe and to Annobon is closer than that of the two smaller islands to each other. Taking all four islands, either with or without endemics, we infer again that San Thomé is more closely related to the three other islands than either of these are to each other. Principe, which is between

Fernando Po and San Thomé, is definitely more closely related to the former than to the latter, although this is between it and the mainland.

Exell (1947) discussed the results that I previously obtained, which differ only very slightly from those above, and summed up "The conclusions . . . obtained by statistical methods agree admirably with the conclusions as to the relationships between the islands which I had already reached in my account of their affinities given in the catalogue."

It should be noted that the object of this particular analysis was not to discover new information about the relations of the floras of these tropical islands, but to test a statistical method against data already well interpreted. If the conclusions reached appeared reasonable, then the method applied to other floras might help to confirm, or to question, theories based on other evidence.

I have referred earlier to the situation where the observed number of species common to two areas might be greater than the number calculated on the assumption of both populations being random samples of a larger one. This can happen under special circumstances. For example, the number of species of Macro-Lepidoptera in Great Britain is approximately 792. Ireland has approximately 550, but of these only one is not found in Britain. Thus there are 549 species common to the two areas, and the addition of the 32,000 square miles of Ireland to the 89,000 square miles of Britain only adds a single species.

When we calculate the number expected to be common by the method used above we find that there should be only 447, so the observed number 549 is 123 per cent of the expected. The explanation is that the faunas cannot be looked upon as random samples from the larger European area, as the Irish fauna has been largely derived directly from that of Britain. With the exception of those species that have reached both countries independently from the south, the species have had to pass through Britain on their way to Ireland.

A relation of over 100 per cent would therefore be an indication that the smaller population had been derived directly from the larger.

Chapter 6

PROBLEMS OF SPECIES, GENERA AND HIGHER GROUPS, AND THEIR BEARING ON CLASSIFICATION

FOR the 200 years since Linnaeus first put forward his ideas taxonomists have been classifying animals and plants by a system based, theoretically, on gradually differing degrees of relationship, which we call species, genera, families, and so on.

The taxonomist believes that by this system he is interpreting something real in nature, and not just making a convenient cataloguing system; but he usually has insufficient evidence, and often is forced to base his conclusions on structures of convenience, instead of what he might consider more fundamental characters which are difficult to observe.

As a result of the work of thousands of systematists we have classifications, in various stages of completeness, for nearly all groups of animals and plants, and it is of interest to enquire if any general pattern or principles emerge as a result of their labours. Such a structure might well throw light on the fundamental natural processes which they are attempting to interpret, and so also on the evolutionary stages and the causes which have contributed to the present momentary balance and pattern.

One general principle, or mathematical pattern, has several times been discussed. In almost every classification that has been proposed the number of genera with only a single species is greater than the number with two, the number with two greater than with three, and so on. If we plot such a classification in the form of a frequency distribution we get a hollow curve, not unlike a hyperbola, which immediately recalls the similar pattern already discussed for the relative abundance of species.

For example, in the catalogue of the Mantidae (Orthoptera) of the world, Kirby (1904) recognized 805 species which he classified into 209 genera, with the distribution of genera of different sizes as shown in Table 49 and Fig. 49.

Willis (1922) made considerable use of this mathematical pattern in his study of "Age and Area", but both he and Chamberlin (1924) who extended the idea, assumed that the distribution had all the properties of a hyperbolic distribution. This was an error, as the sum of a hyperbolic series, both in units and in groups, is always infinite. Chamberlin also made a further error in calculating the fit of his series to a hyperbola.

In 1924, Yule discussed and developed the theories of Willis and obtained a mathematical formula based on certain assumptions about the frequency of mutations or similar processes which result in the start of a new species. One

TABLE 49. *Number of genera with different numbers of species in Kirby's classification of the Mantidae (Orthoptera) of the world*

Sp. per genus	No. of sp. observed	log series	Sp. per genus	No. of sp. observed	log series
1	82 (82)	82.3 (82.3)	12	—	2.1
2	28	36.9	13	2 (52)	1.7 (43.1)
3	27	22.1	15	2	—
4	12 (67)	14.9 (73.9)	16	1	—
5	20	10.7	19	1	—
6	7	8.0	20	1	—
7	4	6.2	21	1	—
8	8	4.8	30	2 (8)	— (9.7)
9	5	3.9			
10	1	3.1		Total species = 805	
11	5	2.5		Total genera = 209	

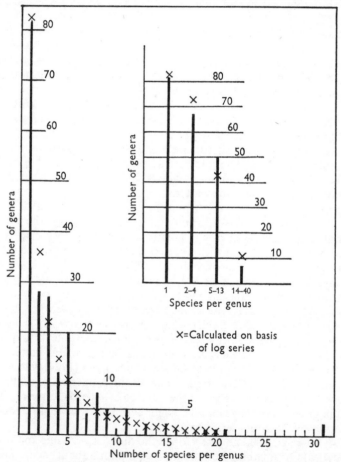

FIG. 49. Genera and species of Mantidae of the world, according to Kirby.

of these assumptions was that in the evolution of species and genera "the chances of the formulation of a new species from another within the same genus is, in any assigned interval of time (an hour, a year, or a century), the same for all the species within the genus considered, and is constant for all time".

This rules out any effect of the number of individuals in the different species, of differing numbers of young produced, and of the length of the life-cycle affecting the frequency of breeding seasons. Thus he assumes that the number of mutations would be unaffected by the number of gene sortings in the reproductive processes. I find it very difficult to accept his assumption.

Yule gives, among other examples, the frequency distribution of genera with different numbers of species in the family Chrysomelidae of the Coleoptera, extracted by Willis from the catalogue by Gemminger and Harold in 1874. The distribution is shown in Table 50, and Fig. 50 shows the typical hollow curve with more genera with one species than at any other level. Yule points out that there is an almost straight-line relation between the log number of species per genus and the log number of genera, the former decreasing as the latter increases. His diagram is reproduced in Fig. 50B. This distribution is the same as that suggested in 1942 by Corbet (see p. 19) for the relation between individuals and species in his collections of butterflies from Malaya. The general formula is $\log S + m \log n = $ a constant, and if $m = 1$ the relation is an hyperbola. The dotted line added to Fig. 50B represents the hyperbola, and it will be seen that it does not agree with the observed data.

TABLE 50. *Number of genera with different numbers of species in a sample of 10,000 species, from Gemminger and Harold's catalogue of the Chrysomelidae of the world*

Sp. per genus	No. of genera	Sp. per genus	No. of genera	Sp. per genus	No. of genera	Sp. per genus	No. of genera	Sp. per genus	No. of genera
1	215	15	8	29	3	45	1	69	1
2	90	16	6	30	3	46	1	71	1
3	38	17	6	32	1	49	2	72	1
4	35	18	3	33	1	50	4	73	1
5	21	19	4	34	1	52	1	74	1
6	16	20	3	35	1	53	1	76	1
7	15	21	4	36	3	56	1	77	1
8	14	22	4	37	1	58	1	79	1
9	5	23	5	38	1	59	1	83	1
10	15	24	4	39	2	62	1	84	3
11	8	25	2	40	2	63	3	87	2
12	9	26	3	41	1	65	1	89	1
13	5	27	1	43	4	66	1	92	2
14	6	28	3	44	1	67	1	93	1

And also single genera with the following number of species:
110, 114, 115, 128, 132, 133, 146, 163, 196, 217, 227, 264, 327, 399, 417, and 681.
Total species = 10,000; total genera = 627

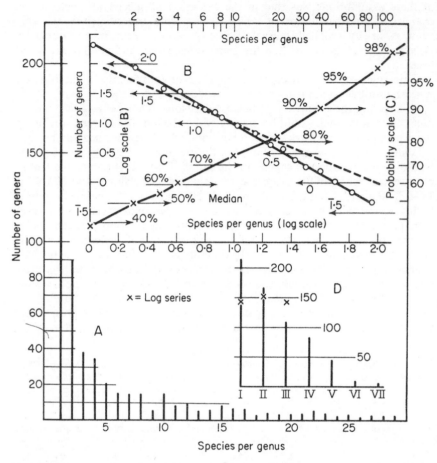

FIG. 50. Genera and species of Chrysomelidae (Coleoptera) of the world according to an extract of 10,000 species by Willis from Gemminger and Harold, 1874 and 1876.

The frequency distribution in these Chrysomelidae will be discussed again below.

It is therefore desirable to study some examples of classification by different authorities and in different groups, and to see whether any general formula, more applicable than the hyperbola, can be found.

The example quoted above of Kirby's classification of the Praying Mantis was first discussed by myself (Williams, 1944, p. 20), and was selected because the classification was almost entirely the work of one man who had personally seen most of the species in the collections at the British Museum. Since we know that there are considerable differences between the opinions of different students even in the same group, it is important to study a classification not complicated by a multiplicity of authorship.

When a logarithmic series is calculated to fit Kirby's 805 species in 209 genera the resulting numbers of genera with different numbers of species is shown in Table 49 and in Fig. 49. It will be seen that the calculated n_1 (genera with one species) is 82.3, while the observed number was 82. The fit for the later terms, $n_2 - n_4$ is not so remarkable, but there is no steady error in one direction, and the total of $n_1 - n_5$ is 169 as compared with the calculated 167. So we see that Kirby's ideas on the classification of the Mantidae of the world (whether correct or incorrect by modern standards) resulted in a pattern of distribution very closely represented by the logarithmic series.

CLASSIFICATION OF THE CHRYSOMELIDAE (COLEOPTERA)

It has been mentioned just above that Yule (1924) used the classification of beetles of the family Chrysomelidae as evidence in favour of some of his theories. Willis had previously extracted a sample including 10,000 species from the catalogue of Gemminger and Harold (1874), and the frequency distribution of genera with different numbers of species is shown in Table 50. The total number of genera was 627, giving an average of 15.95 species per genus. This is a rather high average for modern classifications, but it should be noted that the median value of species per genus is well down in the genera with only three species.

The distribution is shown graphically in Fig. 50A, and the fit to a straight line, as suggested by Yule, with the log number of genera inversely related to the log number of species per genus, is shown in Fig. 50B. The unbroken line drawn to fit as closely as possible the observed values represents the inverse power series $y \times x^{1.27} = 215$. It would give a finite number of genera, but an infinite number of species. To this figure has been added, as a broken line, the condition if the distribution had been hyperbolic.

A log series calculated to fit the total genera and species gives the first term n_1, as 145, instead of 215; the next three terms, $n_2 - n_4$, as 71.5, 46.9 and 34.7 (total 153) instead of the observed 90, 38 and 35 (total 163). But the total of the terms $n_4 - n_{13}$ is 142.9 in the log series and only 108 in the observed (see xxxx in Fig. 50D). Thus the observed distribution has more genera with one and two species, and fewer genera at the higher levels than would be required by the log series.

When the observed genera are grouped into $\times$ 3 classes the totals are: I = 215, II = 163, III = 108, IV = 81, V = 47, VI = 10, and VII = 3. This distribution is shown graphically in Fig. 50D. It might suggest a truncate log normal, but even for that the fit is not good. If, however, the accumulated total of genera is plotted on a probability scale (Fig. 50C) there is a moderately good fit to a straight line. This would imply that, if the distribution were log normal, all the classes below one species per genus were included in this class, and there were no zero terms on the arithmetic scale. It is always unsafe to discuss the fit of a distribution to a log normal when there is no measurable zero class, or when nearly half or more of the total groups are in the n_1 class.

On the whole this set of data within its range fits better to Yule's exponential formula, but with such a large number of species it is unlikely that the classification is uniformly critical, and it will almost certainly be composite, different parts having been evolved by different minds with different conceptions of generic divisions.

GENERA AND SPECIES OF THE BUTTERFLIES OF AUSTRALIA

Waterhouse (1932) recognized 351 species which he classified into 111 genera. The number of genera with different numbers of species are shown in Table 51, together with the earlier terms of a log series calculated to fit the total genera and species.

TABLE 51. *Genera and species of butterflies in Australia, from the classification of Waterhouse*

Sp. per genus	No. of sp. observed	No. of sp. log series	Sp. per genus	No. of sp. observed	No. of sp. log series
1	52	49.0	8	3	2.12
2	21	21.1	9	—	1.61
3	7	13.5	10	3 (11)	1.25 (11.65)
4	7	7.8	11	2	
5	8 (95)	5.4 (96.8)	12	—	
6	3	3.84	13	2	
7	2	2.83	14	1	

Total: 351 species in 111 genera

It will be seen that the observed number of genera with a single species is fifty-two, slightly above the calculated forty-nine. For all genera with up to five species the totals are ninety-five observed and 96.8 calculated; and for $n_6 - n_{10}$ the numbers are eleven and 11.65. The fit is, for biological data, quite as good as could be expected.

CLASSIFICATION OF THE FLEAS OF THE WORLD

Da Costa Lima and Hathaway (1946) published a list of the species, genera and higher groups of the Siphonaptera of the world in which they recognized ten families, thirty sub-families, 175 genera and 1012 species. The frequency distribution of the units and groups at different levels of classification are shown in Table 52, together with the earlier terms of calculated log series.

The fit to the calculated series is moderately good at all levels, but, as usual, the observed number of genera with one species is higher. I have previously suggested that this may be due partly to a quite natural greater interest of the systematist in a species belonging to a new genus, and partly to the occasional erection of a new genus without justification. In the case of the genera and species the excess in the observed n_1 is more than offset by a shortage in n_2.

TABLE 52. *Genera and species of the fleas (Aphaniptera) of the world, as classified by da Costa Lima and Hathaway*

Units per group	Genera and species		Sub-families and genera		Families and sub-families	
	obs.	log series	obs.	log series	obs.	log series
1	60 (60)	57.8 (57.8)	8 (8)	10.7 (10.7)	3 (3)	4.5 (4.5)
2	24	27.2	6	5.0	2	1.9
3	16	17.2	2	3.5	2	1.1
4	8 (48)	12.1 (56.5)	5 (13)	2.0 (10.5)	— (4)	0.7 (3.7)
5	9		1		1	
6	10		—		2	
7	10		1			
8	11		1			
9	2		1			
10	1		2			
11	5		—			
12	2		—			
13	1 (51)		1 (7)			

and at: 14, 15, 16, 17, 18, 19 (2), 21, 22, 23, 26, 32, 41, 47, 53, and 56.   and at 27 and 40.

Total: 10 families; 30 sub-families; 175 genera and 1011 species.

At both the sub-family genus and the family–sub-family level of classification the observed n_1 is lower than the calculated, the difference being again offset by a reverse effect in $n_2 - n_4$.

Thus at all levels of classification there is evidence of a mathematical pattern of frequency distribution which is closely represented by the log series.

GENERIC SIZE DISTRIBUTION OF LEAFHOPPERS IN FINLAND

In 1950, Kontkanen published a study of leafhoppers (Homoptera, Auchenorrhyncha) in the province of North Karelia, in Finland.

Some of his data on the frequency of genera with different numbers of species, are summarized in Table 53. Kontkanen states that the known fauna of N. Karelia, for this group of insects, is 172 species. Table 53A and Fig. 50A give the generic distribution of all the species found during his study in the area specially examined. The 159 species belong to seventy-two genera, which gives an average of 2.21 species per genus; a generic diversity, from the log series, of 50; and an estimated number of monotypic genera (n_1) of 38.1 as compared with the observed 41.

Table 53B and Fig. 51B give a summary of his observations on samples from thirty different biotypes combined. The total of eighty-seven species in fifty-two genera gives an average of 1.67 species per genus; a diversity of 53.4; and an estimated n_1 of 33.1 as compared with the observed 34. First we should notice the very close fit of the observed to the calculated number of

TABLE 53. *Genera and species of leafhoppers (Homo-ptera) in different areas in Finland; from Kontkanen*

Sp. per genus	A Total area	B Thirty biotypes
1	41	34
2	12	11
3	7	3
4	6	2
5	3	1
6	—	—
7	1	—
8	—	—
9	1	1
18	1	—
Total genera	72	52
Total species	159	87
Diversity	50	54
n_1 (from log series)	38.1	33.1

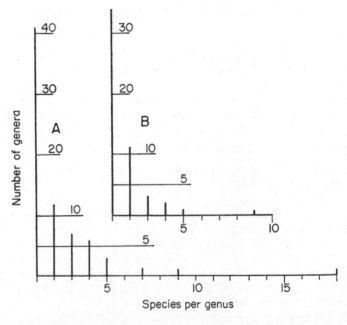

FIG. 51. Frequency distribution of genera of leafhoppers with different numbers of species in North Karelia, Finland. A = total fauna of the area with 159 recorded species; B = combined samples from thirty different biotypes specially studied, with a total of eighty-seven species (Kontkanen, 1950).

monotypic genera, especially in the second series B. Secondly that the increase in the generic diversity, from 50 to 53, although small, is to be expected from the fact that the samples were taken from a specially selected variety of habitats, with almost certainly a greater diversity of environment in relation to total area than in the larger but more general sample (see also p. 161).

CLASSIFICATION OF THE CRABRONIDAE (HYMENOPTERA) OF THE WORLD

Dehalu and Leclercq (1951) discussed the classification by the latter author of the Crabronidae of the world in relation to the logarithmic series. He recognized 700 species which he placed in thirty-one genera. The average number of species per genus, 22.6, is unusually high for any classification; in fact the average number of species per sub-genus (of which he recognized eighty-two) is, at 8.5 much nearer the usual average for genera.

Table 54 and Fig. 52 show the frequency distribution of species in genera and sub-genera, together with the early terms of a log series calculated by

TABLE 54. *Genera and species of the Crabronidae (Hymenoptera) of the world, as classified by Leclercq*

Sp. per group	No. of genera obs.	genera log series	No. of sub-genera obs.	sub-genera log series	Sp. per group	No. of genera obs.	No. of sub-genera obs.
1	4 (4)	6.6 (6.6)	15 (15)	23.3 (23.3)	21	—	2
2	5	3.3	11	11.3	22	—	2
3	2	2.1	7	7.2	30	1	—
4	— (7)	1.6 (7.0)	10 (28)	5.3 (23.8)	33	1	—
5	3	1.27	9	4.07	34	—	—
6	3	1.05	4	3.28	36	1	—
7	—	0.88	1	2.72	37	— (6)	1 (11)
8	2	0.77	3	2.30	43	—	1
9	—	0.68	2	1.98	49	—	1
10	1	0.61	2	1.72	50	—	1
11	—	0.54	3	1.51	53	1	1
12	—	0.50	—	1.37	61	1	—
13	— (9)	0.46 (6.8)	— (24)	1.23 (20.2)	80	1 (2)	— (4)
15	1		3		133	1	—
16	1		1		144	1	—
18	—		1				
20	1		—				

Total species 700   Total sub-genera 82   Total genera 31

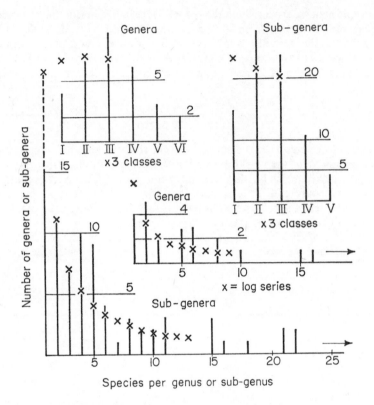

Fig. 52. Genera, sub-genera and species of Crabronidae (Hymenoptera) of the world, according to Leclercq (1954).

Dehalu and Leclercq. It will be noted that an unusual feature of the classification is that there are more genera with two species than with one, although the difference is small. In the sub-genera the difference is reversed but again the observed n_1 is low, and the n_2 is high for the log series.

Dehalu concluded that the observed number of genera with up to nine species—nineteen—is so close to the 18.2 required by the log series that it is difficult to believe that the result is accidental. An examination of the distribution in × 3 log classes shows, however, a distinctly poor fit to the log series for the first three classes up to thirteen species per genus. On the other hand the fit to a straight line, indicating a log-normal type of distribution, in Fig. 53 is distinctly close, although still irregular.

This example is discussed because, in spite of the published opinion of Dehalu and Leclercq that it resembles a log series, it appears to be a case—exceptional in classification patterns—where the log normal gives a better estimate than the log series.

GENERA AND SPECIES OF BRITISH NEUROPTERA

Killington (1937) reviewed the classification of the British Neuroptera, and recognized fifty-three species in eighteen genera and in five families. The number of genera with increasing numbers of species, together with the early terms of a log series, are as follows:

		1	2	3	4
Species per genus					
Number ⎰Observed		9	3	2	1
of species ⎱Log series		8.4	3.53	1.98	1.61

Also single genera at 5, 11 and 12 species

Although the total numbers of genera and species are small, the fit to the log-series distribution is very close.

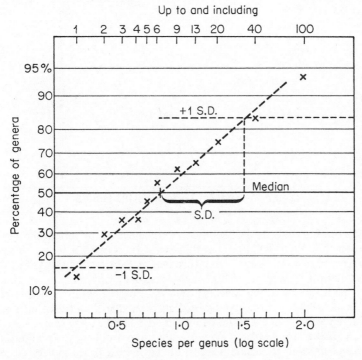

FIG. 53. Crabronidae of the world, as in Fig. 51, in relation to the log-normal distribution.

SPECIES, GENERA AND FAMILIES IN THE BRITISH FISHES

Jenkins (1936) published a list of the Fishes of Britain in which he recognized 374 species, 232 genera and 103 families. Thus the average number of species per genus is 1.61, of genera per family is 2.25 and of species per family 3.63. The frequency distribution of groups with different numbers of units in each of these classification relationships is shown in Table 55. There is also shown the first five terms of log series calculated to fit the particular number of units and groups.

TABLE 55. *Genera and species of British fishes as classified by Jenkins in 1936*

Units per group	Families and genera obs.	log series	Genera and species obs.	log series	Families and species obs.	log series
1	57 (57)	53.8	169 (169)	153	39 (39)	41.8
2	20	20.7	39	45.2	24	18.6
3	9	10.6	12	17.8	10	11.0
4	5 (34)	6.1	5 (56)	8.0	7 (41)	7.4
5	4	3.8	1	3.7	3	5.2
Total 1–5	95	95.0	226	227.7	84	84.0
6	2		1		4	
7	2		—		4	
8	2		—		1	
9	—		2		1	
10	—		1		2	
11	—		—		1	
12	—		—		2	
13	2 (12)		— (5)		— (19)	
14	—		—		1	
15	—		1		1	
17	—		1		—	
26	—		—		1	
30	—		—		1	

Total: 374 species; 232 genera; 103 families

It will be seen that for the classification of species into genera the observed n_1 is larger than the calculated, but this excess is offset by $n_2 - n_5$, so that the total observed and calculated up to this level are almost the same. The family –genus relation is similar with a slight excess of observed n_1, in this case exactly made up by a shortage in $n_2 - n_5$. For the families with different numbers of species, the observed n_1 is slightly smaller than the calculated, but this difference is again exactly balanced by the small excess in $n_2 - n_5$.

There is therefore strong evidence that the frequency distributions of groups and units in this classification of the British fishes is very close to a log-series distribution.

THE CLASSIFICATION OF THE BIRDS OF THE WORLD
ACCORDING TO LINNAEUS IN 1758–59

It is interesting to compare the patterns in modern classifications with those of the earliest systematists. Linnaeus, in the 10th edition of his *Systema Naturae*, published in 1758–9, recognized 553 species of birds of the world, which he classified into sixty-three genera and six orders.

The frequency distribution of genera with different numbers of species is shown in Table 56. Already the pattern with more smaller, and fewer large genera is apparent, but the number of monotypic genera is not so dominant as it became later. The number of genera in the × 3 classes of abundance of species is also shown in the table, and when the accumulated totals in these

TABLE 56. *Classification of the birds into genera and species by Linnaeus in 1758–59*

Sp. per genus	Genera	Sp. per genus	Genera	Sp. per genus	Genera
1	9 (9)	9	1	19	1
2	8	10	1	22	1
3	4	11	2	26	1
4	9 (21)	12	2	31	1
5	4	13	3 (21)	32	1
6	4	14	2	34	1
7	2	16	1	37	1
8	2	18	1	39	1 (12)

Total: 63 genera; 553 species

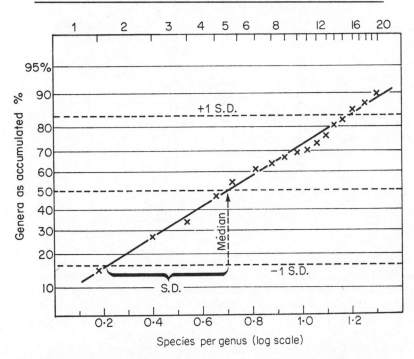

FIG. 54. Relation of the classification of birds into genera and species by Linnaeus, and the log-normal distribution.

groups are plotted as percentages of the total genera against a probability scale, as in Fig. 54, the result is extremely close to a straight line. So Linnaeus's ideas of classification nearly 200 years ago have quite unconsciously produced a pattern very close to a log-normal distribution.

PASSERINE BIRDS OF THE WORLD AS CLASSIFIED BY SHARPE

In R. Bowdler Sharpe's "Handbook of the Genera and Species of Birds" (1899–1909) the greater part of three volumes is devoted to the passerine order. In this he recognizes 10,138 species in 1454 genera, with an average of 6.97 species per genus, as compared with the 8.8 per genus for Linnaeus.

TABLE 57. *Genera and species of the birds of the world according to R. Bowdler Sharpe 1899–1909*

Sp. per genus	No. of species	Sp. per genus	No. of species	Sp. per genus	No. of species
1	455 (455)	18	8	35	5
2	227	19	14	36	2
3	137	20	9	38	1
4	90 (454)	21	6	39	1
5	67	22	11	40	1 (164)
6	39	23	5	41	3
7	45	24	8	42	2
8	43	25	5	43	1
9	36	26	8	45	1
10	40	27	5	46	3
11	29	28	4	47	1
12	28	29	4	49	1
13	20 (347)	30	2	50	1
14	13	31	3	51	3
15	18	32	3	52	4
16	8	33	1	53	1
17	16	34	3	54	1

Also single genera with: 55, 56, 58, 60, 66, 72, 74, 84, 88, 99, 114, and 157 species.
Total: 10,138 species in 1454 genera

Table 57 shows the number of genera with different numbers of species, together with the numbers grouped into × 3 classes. It will be seen that with the arithmetic classes there is the familiar hollow-curve distribution with almost exactly twice as many genera with one species as with two. With the × 3 classes the first (n_1) and the second ($n_2 - n_4$) are almost exactly equal, suggesting a truncate log normal just including the peak.

A log series calculated to fit the total genera and species gives the first five terms, with the observed values in brackets, as: 444.6 (455), 212.5 (227), 135.5 (137), 97.1 (90), and 74.2 (67). So the observed total of genera with up to five is 976 and that calculated on the assumption of a log series is 974.3. One could not expect a much closer correspondence.

When the accumulated percentage of genera is plotted on a probability scale, as in Fig. 55, the result is a very close fit to a straight line, indicating a log-normal distribution with a median at 0.46 and a S.D. of 0.56 on the log scale, or 2.88 $\times / \div$ 3.63 species per genus on the arithmetic scale.

The close fit of the data to both these distributions indicates the difficulty of distinguishing a very truncate log normal from a log series.

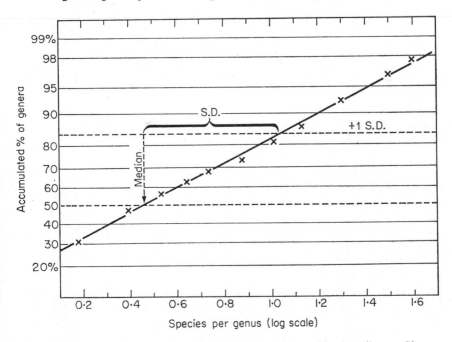

FIG. 55. Genera and species of passerine birds of the world, according to Sharpe (1899–1909), in relation to the log-normal distribution.

THE PATTERN OF CLASSIFICATION AS ILLUSTRATED BY THE BRITISH BIRDS

The birds of Great Britain and Ireland are probably as well known as those of any similar area in the world, and the British Ornithological Union check list, published in 1952, recognizes nineteen orders, fifty-nine families, 191 genera, 426 species and 528 sub-species as having been recorded in the area.

Table 58 and Fig. 56 show the frequency distribution of groups with different numbers of units at each classification level, e.g. the number of orders with different numbers of families, the families with different numbers of genera and so on. The table also shows estimates of the value of the number of groups with one unit (n_1) and for the generic diversity, calculated on the

TABLE 58. *Frequency distribution of families, genera and species of birds in the British fauna, showing the pattern of classification in the British Ornithological Union's List of 1952*

Units per group	Orders and families	Families and genera			Genera and species			Species and sub-species		
	All	Non-Pass.	Passer-ine	All	Non-Pass.	Passer-ine	All	Non-Pass.	Passer-ine	All
1	9	17	10	27	79	36	115	236	115	351
2	5	8	3	11	25	9	34	25	29	54
3	2	3	2	5	6	3	9	4	13	17
4	1	3	—	3	9	2	11	—	2	2
5	—	1	1	2	2	3	5	—	2	2
6	—	1	1	2	1	2	3	—	—	—
7	—	1	—	1	2	3	5	—	—	—
8	—	—	1	1	1	—	1	—	—	—
9	—	3	1	4	1	1	2	—	—	—
10	1	—	—	—	—	2	2	—	—	—
11	—	—	—	—	1	—	1	—	—	—
12	—	—	1	1	2	—	2	—	—	—
13	—	—	—	—	—	1	1	—	—	—
14	—	1	—	1	—	—	—	—	—	—
16	—	1	—	1	—	—	—	—	—	—
20	1	—	—	—	—	—	—	—	—	—
Total groups	19	39	20	59	129	62	191	265	161	426
Total units	59	129	62	191	265	161	426	298	230	528
Unit per group	3.1	3.31	3.10	3.24	2.05	2.60	2.23	1.12	1.43	1.24
Estimated n_1	8.1	16.8	8.9	25.8	71.6	29.9	100	238	117	354
Diversity	10	19	10	30	98	37	131	1192	239	1073

assumption that the frequency distributions can be closely represented by the logarithmic series. The frequency distributions and estimated parameters are shown separately for the passerine birds (one order only) and the non-passerines, which include eighteen orders.

It will be seen that in every case the frequency distribution is of the familiar hollow-curve type with more groups with a single unit than at any other level. The resemblance to the log series is supported by the close, and sometimes very close, fit between the estimated and observed values of n_1. For example, for the species and sub-species of non-passerine birds the estimated value of n_1 is 238 and the observed 236, while for the families and genera it is 16.8 and the observed 17. In the passerine birds the estimated number of species with one sub-species in 117 and the observed number was 115. When there is a difference it is nearly always that the calculated is a little below the observed.

Since the different groups studied are not, in this case, random samples by units of a single larger population, the comparison of the diversities is not of any great value, except that one can note the expected greater diversity in the non-passerine groups with the much larger number of orders. The average number of units per group is highest in the family–genus relation and lowest in the species–subspecies relation.

In the comparison of passerine birds with the others, the number of units per group is higher in the latter at the family–genus level, but lower in the genus–species and species–sub-species relationship.

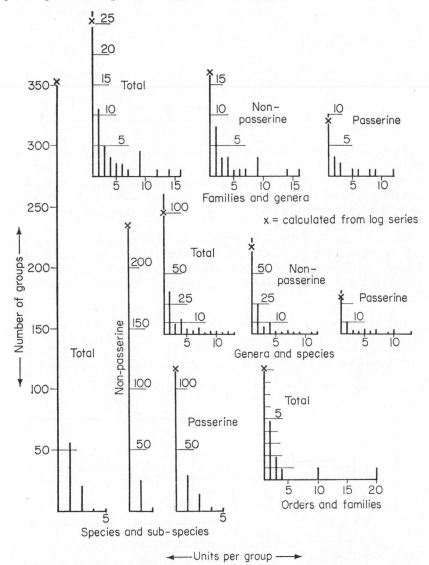

FIG. 56. Patterns of classification in the British birds, showing the frequency distribution for species and sub-species, genera and species, families and genera, and orders and families. In each case (except the orders and families) separate diagrams are given for the total recorded fauna, and for the passerine and non-passerine birds separately.

TABLE 59. *Genera and species of the British birds in the B.O.U. list of 1952,*
with the corresponding values calculated from the logarithmic series

Sp. per genus	No. of genera obs.	calculated	Sp. per genus	No. of genera obs.	calculated
1	115 (115)	100 (100)	8	1	1.92
2	34	38	9	2	1.30
3	9	19.5	10	2	0.90
4	11 (54)	11.2 (68.7)	11	1	0.62
5	5	6.85	12	2	0.44
6	3	4.37	13	1 (22)	0.31 (19.77)
7	5	3.06	over 13	—	2.5

To study the genera and species relation in more detail a log series has been calculated to fit 426 species in 191 genera, and the first thirteen terms are shown in Table 59 together with the numbers found in the check list. It will be seen that, as usual, there are more observed genera with a single species than the calculated number, but both n_2 and n_3 show the reverse, and from n_3 on the sums are almost identical.

For comparison with Fig. 54, showing the fit of Linnaeus's classification to a log normal, Fig. 57 shows a similar treatment for the more recent classification. It will be seen that the accumulated totals form a moderate fit to a straight line, but as the first term (monotypic genera) include over half of the

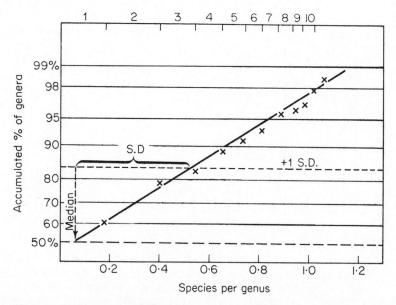

FIG. 57. Genera and species of British birds, from the British Ornithological Union's 1952 list, in relation to the log-normal distribution.

total genera, it is doubtful if any significant deduction can be made. In a truncate distribution of this type it is possible that the upper portion might closely resemble both a log series and part of a log normal.

CLASSIFICATION OF THE BRITISH FLORA IN 1912

The 1912 edition of Bentham and Hooker's "British Flora" recognized 1251 species in 479 genera and in eighty-nine natural orders, now more usually referred to as families. The average number of species per genus is 2.61 and of genera per order is 5.38. The frequency of genera with different numbers of species, and of orders with different numbers of genera, are shown in Table 60, together with the log series calculated for the first thirteen terms.

TABLE 60. *Orders, genera and species of the British flora according to Bentham and Hooker in 1912*

Units per group	Genera and species obs.	log series	Orders and genera obs.	log series
1	256 (256)	231.3 (231.3)	32 (32)	29.3 (29.3)
2	84	94.3	17	13.8
3	51	51.3	9	8.60
4	22 (157)	31.3 (176.9)	8 (34)	6.06 (28.5)
5	21	20.4	3	4.55
6	8	13.9	2	3.56
7	5	9.69	1	2.87
8	5	6.91	—	2.35
9	5	5.01	3	1.96
10	6	3.67	1	1.66
11	4	2.72	1	1.42
12	3	2.03	—	1.22
13	1 (58)	1.53 (65.9)	1 (12)	1.06 (20.65)
Total 1–13	471	474.1	78	78.5
	also at 14, 15 (3), 16, 17, 21 and 47.		also at 14 (2), 16 (2), 17, 18 (2), 27, 36, 41 and 42.	

Total: 1251 species; 479 genera; and 89 orders

It will be seen that at both levels of classification the observed number of groups with one unit exceeds the calculated by about 10 per cent. In the species–genus relation this excess is more than levelled out by a deficiency in most of the terms from n_2 to n_9, after which there is again a slight excess of observed over calculated. So the smallest genera and the larger ones are slightly more frequent, and those of intermediate size slightly less frequent than required by the log series.

In the genus–order relation the sequence is different, as there are more than the calculated number of orders in n_1 to n_4, and after this a reversal to fewer. Groups n_1—n_4 have a total of 66 observed orders against 57.8 calculated, but by n_{13} the sums are almost identical at 78 observed against 78.5 calculated.

These results are not conclusive, but do not completely rule out the log series. On the other hand, graphical tests with a probability scale (Fig. 58) indicate a possible fit of the log normal to the species–genus distribution, but considerable departures from this for the genus–order relation, particularly in the larger orders.

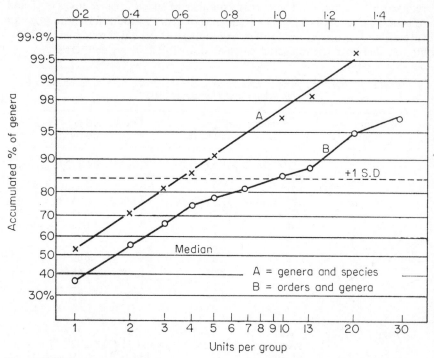

Fig. 58. Orders, genera and species of the British flowering plants according to Bentham and Hooker (1912) in relation to the log-normal distribution.

CLASSIFICATION OF BRITISH NATIVE FLOWERING PLANTS IN 1952

In 1952, Clapham, Tutin and Warburg published a British flora from which the details are taken for Table 61 for the sizes of genera of our native flowering plants. A total of 1537 species are classified into 525 genera, with an average of 2.93 species per genus.

The table also shows the first thirteen terms of a log series calculated to fit the total genera and species. From this it will be seen that the observed number of genera with only one species is considerably above the calculated, the difference being forty genera or 15 per cent. In the terms n_2 — n_{10}, how-

TABLE 61. *Genera and species of the British flowering plants, according to Clapham, Tutin and Warburg in 1952*

Sp. per genus	No. of genera obs.	No. of genera log series	Sp. per genus	No. of genera
1	278 (278)	237.9 (237.9)	14	3
2	94	100.6	15	1
3	48	56.7	17	1
4	31 (173)	35.9 (193.1)	18	2
5	18	24.3	19	1
6	9	17.1	23	1
7	6	12.4	24	2
8	7	9.2	25	1
9	4	6.9	40	— (12)
10	5	5.2	41	1
11	4	4.0	77	1
12	3	3.1		
13	4 (60)	2.4 (84.7)		
Total 1–13	511	515.7		

Total: 1537 species in 525 genera

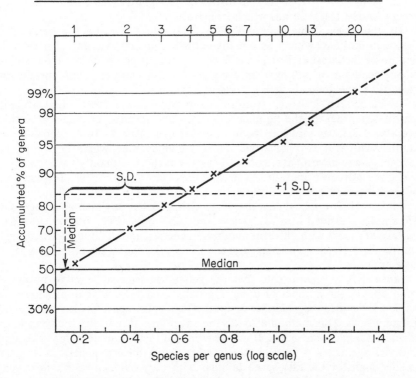

FIG. 59. Genera and species of the British flowering plants according to Clapham, Tutin and Warburg in 1952, in relation to the log-normal distribution.

ever, the calculated are all above the observed. In the sum up to n_7 the values are nearly equal, and by n_{13} the sum of the calculated is greater. Thus the observed distribution has more genera with only one, and with more than fourteen species, but fewer in the range n_2 to n_{14}, than is suggested by the log series.

A graphical examination of the fit to the log normal, shown in Fig. 59, from the accumulated percentages of genera with increasing numbers of species, indicates a close fit to a straight line. As, however, the number of genera with one species is over half the total genera, the relation is not easy to interpret. It suggests a log normal with a median on the log scale at 0.14 (1.38 species per genus) and a S.D. of 0.5.

THE RELATION BETWEEN THE SIZE OF A FAMILY AND THE SIZE OF THE GENERA CONTAINED IN IT

Willis (1922), discussing the relation between species, genera and families from the point of view of his theory of "Age and Area", writes (p. 237): "If one takes a number of species per family in the British flora, one finds it to increase steadily with the number of genera. There are no breaks, as one would be inclined to expect." He instances the figures given here in Table 62, except for the last column, which I have added.

He continues: "One may find here, as elsewhere, that (as a general rule) the small families, which, as already explained under 'size and species' will tend to be the latest arrivals, have fewer species per genus. While the families of one genus in Britain have an average of 2.2 species per genus, those with more than one genus have a general average of 3.3."

It will be seen, however, from the figures added to Table 62, that while it may be true that all the families with over one species per genus have this average of 3.3, the sequence within the classes is very far from regular. The families with only two genera have a very high average, while those with six to eight genera have an average nearly as low as those with only a single genus.

To get more information I have examined below data for the classification of the fishes and birds of Britain.

Jenkins (1936) published a list of the British fishes in which he recognized 374 species classified into 232 genera and 103 families. The frequency distributions of genera with different numbers of species, and of families with different numbers of genera and species, as in Table 63, shows the familiar type of distribution.

Since it is a general rule that the geometric scale appears to be more applicable to the number of units in groups, I have calculated, as shown in Table 63, the mean *log* number of species per genus separately for families of different sizes. This method also has the advantage of reducing the swamping effect of a single large genus. It will be seen that there is no definite trend, and that the average log for the two families with thirteen genera is a little smaller than for the fifty-seven families with only one genus. As, however, the number of small families is high compared with the number of larger ones, I have recalculated the mean logs with the families reclassified into

TABLE 62. *The relation between the size of family and the mean number of species per genus within the family, from the British flora, data from Willis*

	Arithmetic mean	
Genera per family	Species per family	Species per genus
1	2.2	2.2
2	8.3	4.15
3	10.4	3.4
4	12.3	3.1
5, 6, 7	15.0	2.5
8, 9, 10	40	4.4
over 10	73	—

TABLE 63. *Relation between the size of family and the mean size of the genera within the family in the British fishes*

Genera per family	No. of genera	Mean log species per genus
1	57	0.13 (0.13)
2	20	0.105
3	9	0.12
4	5	0.24 (0.13)
5	4	0.115
6	2	0.085
7	2	0.04
8	2	0.07
13	2	0.11 (0.09)
Total of all 103 families		0.127

those with one, two to four, five to thirteen genera. Here the mean log is the same in the first two groups, and the third is slightly smaller. It is doubtful if the difference is significant, but it does not support Willis's theory.

From the British Ornithological Union's Check List of British Birds (1952) I have extracted the information given in Table 64 on the classification pattern of species, genera and families.

The families were classified in ascending order of the number of genera they contained, and for each class the average number of species per genus has been calculated by two methods. The first is by adding together the total genera and the total species in all families in the same class, and getting an arithmetic mean of the number of species per genus; the second by converting the number of species in each genus to a log scale and finding the mean log species per genus for each family. The average for all the families in one class is then obtained by adding their separate averages together and dividing by

TABLE 64. *Relation between the number of genera per family and the mean number of species per genus in the British birds*

Genera per family	No. of families	Arithmetic mean species per genus in the class	Mean log species per genus in the class	
1	26	1.73	0.15	
2	11	1.77	0.17	
3	5	1.60	0.13	
4	3	1.91	0.14	
5	2	1.70	0.145	
6	2	1.50	0.14	
7	1	2.29*	0.24*	
8	1	4.75*	0.51*	
9	4	2.33	0.22	
12	1	2.75*	0.23*	
14	1	2.93*	0.22*	
16	1	2.63*	0.31*	
				Antilog.
1	26	1.73	0.15	1.41
2–4	19	1.75	0.16	1.44
5–13	11	2.42	0.22	1.66
over 13	2	2.77	0.265	1.84

* Based on only a single family and so subject to considerable error.

the number of families. Since all families in the same class have the same number of families, this average is valid.

The table also shows the results when the classes are grouped to show the values for families with one, two to four, five to thirteen and over thirteen genera.

The result of this is more in support of Willis's theory than either his own example or that given above for the fishes. There is an irregular but steady increase in species per genus in the different-sized families, except for the value of 4.75 arithmetic mean (0.51 log scale) for the class with eight genera, but this (together with the other classes marked *) is for a single family only, and so subject to considerable error. The geometric × 3 classes show a steady and even rise in size of genera.

Willis's theories may be correct, but it is important that the data on which this, or any other theory, is based should be sound statistically and should cover a wide field of examples.

THE INTERPRETATION OF CLASSIFICATION BY DIFFERENT TAXONOMISTS

All classifications are the outcome of human minds attempting to interpret, usually from insufficient evidence, the relationships of species or genera in their evolutionary development over long periods of time. No two taxonomists will come to exactly the same conclusions, but each believes that what he calls genera are groups of species which have had some common origin in the not too distant past. Families are groups of genera whose common origin is still further back. There are, however, so far as we know, no great discontinuities in evolution and so the dividing lines are often ill defined and their exact position a matter of opinion.

Some taxonomists will allow genera to be separated on characters which others consider too small or too recent in origin to be a justification for the separation. Other taxonomists will allow the division of genera only on characters which have a much more distant origin, and are thus more "fundamental". For a given number of species the former will allow more genera than the latter. The extremes are popularly known as the "splitters" and the "lumpers". The splitter produces a classification with a high generic diversity; the lumper one with a low diversity.

It will be seen, for example, that Kirby (Table 49) classified the 805 species of Mantidae, that he recognized for the whole world, into 209 genera with a mathematical frequency distribution closely resembling a logarithmic series, and with a generic diversity of 92.

One might be inclined, on this account, to say that the logarithmic series was a "correct" interpretation of the structure of the genera–species pattern in the Mantidae, and that Kirby had given a correct interpretation of this structure. It would, however, probably be better to say that his interpretation was consistent rather than it was "correct" in the sense that any other classification would be incorrect.

Had Kirby been a splitter, and so willing to accept smaller differences as a basis for the erection of genera, he might have classified his 800-odd species into 400 genera; had he been a lumper he might have recognized only 100 genera. For each of these alternatives a logarithmic series can be calculated as follows:

Log series for 809 species in 400 genera:
225, 81, 43, 23.5, 15 . . .
Log series for 809 species in 100 genera:
29, 14, 9, 6.5, 5 . . .

It is probable that, had our hypothetical "Kirby" been equally consistent in each case in the application of his principles, his classification would have conformed to one or other of these series. In other words, all three classifications would be equally sound in so far as they express the evolutionary relationships.

What then can we conclude about right and wrong classifications and the validity of genera, species and higher groups?

My own opinion used to be that "species", as described by taxonomists, were about 90 per cent real biological interpretations and about 10 per cent matters of convenience. Genera, on the other hand, I considered to be about 10 per cent real and 90 per cent matters of convenience. When, however, it appeared that the same mathematical pattern fitted the grouping of individuals into species, and of species into genera, I began to consider that genera and families might be as real (or as unreal) as individuals and species.

When, later, we found that one mathematical pattern, by the alteration of the constants only, could fit a number of different interpretations of the same data, a new problem arose; and we must ask if the splitter and the lumper may equally truthfully interpret what has happened in the process of evolution.

It is not possible to support such an hypothesis by direct evolutionary evidence, but it can be shown that it is not impossible theoretically.

In Fig. 60 I have shown an imaginary history in time of the closeness of relation of twenty-three species supposed to be existing at the present era. They have branched off in the past from the common stock in the way that species and genera are generally assumed to have done.

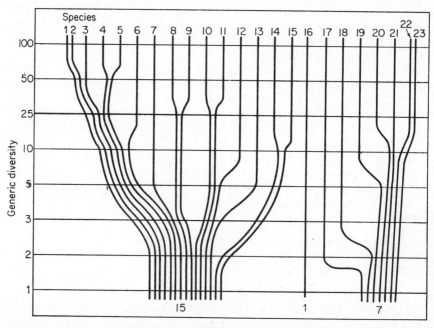

FIG. 60. Imaginary evolution of twenty-three species and their divergences into genera in past time, resulting in a log-series frequency distribution of genera with different numbers of species at different periods of time, with different measures of generic diversity.

If the splitter takes these species and is prepared to accept as generic differences any divergences that might have occurred in quite recent times, he would classify them into nineteen genera with one species each, and two genera with two each, as in the top transverse line. This is an approximation to a log series with a generic diversity of 100 (see Table 65).

The man who is less of a splitter probes more deeply into the past for his generic separation and may end up (second transverse line) with one genus with three species, two with two, and sixteen with one species. This approximates to a log series with a diversity of 50.

Similarly, the taxonomist requiring larger and more fundamental differences before he will accept a genus, will unconsciously probe still deeper into evolutionary history, till perhaps the extreme lumper will only allow three genera in all, with one, seven, and fifteen species respectively. This is an approximation to a log series with a diversity of 1.

Species and genera that have ceased to exist during this period of evolution have not been taken into account, as the problem concerns only the classification of existing species.

I do not wish to press too seriously this purely hypothetical pedigree, or the suggestion that the log series is the only pattern that should be considered. It is, however, possible that any of the half-dozen alternative classifications given above would be an equally "true" account of the evolutionary relations.

Kipling has written

There are nine and sixty ways of constructing tribal lays,
And every single one of them is right!

Perhaps, if we replace "right" by "consistent with the evidence", the same may be true of human efforts at classifying animals and plants, and that "consistency" is all that we can hope to achieve.

TABLE 65. *Twenty-three species divided, theoretically, into different numbers of genera, with different generic diversities, each arrangement corresponding to a different logarithmic series*

No. of genera	Generic diversity	No. of genera with—species									
		1	2	3	4	5	6	7	9	13	15
21	100	19	2	—							
19	50	16	2	1							
16	25	12	3	—	—	1					
12	10	7	3	—	1	—	1				
9	5	4	2	1	—	1	—	1			
7	3	3	1	—	1	1	—	—	1		
5	2	2	1	—	—	—	1	—	—	1	
3	1	1	—	—	—	—	—	1	—	—	1

What would be "wrong" would be to delve deeply at one place and less deeply at another; to be at one point a splitter and at another a lumper, perhaps to support one's own pet theory or to dispose of another's. Thus, it may happen, that a composite publication by several authors could be, generally, less reliable than one individual's work. It was for this reason that I chose for my own starting-point in this study the early work of Kirby, as it was nearly all the work of one man, and nearly all the species were represented in the collections that he studied so that he could come to his own conclusions.

Looking back on my own early efforts as a systematist, from this point of view, I can see that I had no innate conception of what I considered to be a genus, and overemphasized mere convenience. If a genus was particularly large I tried to find characters that would enable it to be divided. Thus I had different standards for small and for large genera. The mathematical pattern shows that there must be large as well as small genera, their relative frequency depending on the generic diversity, which is determined by the consistent conception in the mind of the classifier.

I think that insect taxonomy benefited when I discontinued work in this field.

Chapter 7

DIVERSITY AS A MEASURABLE CHARACTER OF A POPULATION

If one goes into a natural forest in a cold temperate climate such as Northern Europe and selects at random two trees, the chances are high that both will belong to the same species, because in such an environment the vegetation is undiversified. If one makes the same experiment in a tropical forest, it may be necessary to select quite a number of pairs before getting two of the same kind: here the vegetation is highly diverse.

If one collected a few thousand mosquitoes in the far north, it is likely that only a few species would be represented, but in the tropics forty or fifty species might easily be found in a sample of the same size.

If, by some form of trapping, such as a light trap, one collected a hundred moths in Britain in early spring, only a few species would be represented; if the same number were collected in midsummer, there might easily be four or five times as many species. In summer the population is more diversified than at the colder times of the year.

All these are examples of the effects of different specific diversification in populations, and the character is of particular interest to ecologists because it belongs to the population and not to any of the species which compose it.

One of the simplest approaches to the measurement of diversity, first suggested by Simpson in 1949, follows from the remarks in the first paragraph. It is a calculation of the number of pairs that would have to be selected at random from a particular population in order to give an even chance of getting one pair with both individuals belonging to the same species. If we know the make-up of the whole population, the calculation is comparatively simple, as follows:

Suppose a population consists of a certain total number of individuals, including all species, which we call N. Then the first of a pair of individuals selected at random might be any of the N, and the second any of the $(N-1)$ left after the first had been taken. Thus all possible pairs can be selected in $N(N-1)$ ways; but since, for our purposes, the order in which the two are selected is immaterial (that is to say that individual x followed by individual y is the same as y followed by x), the number of *different* pairs is half the above, or $N(N-1)/2$.

Now, if within the population there are species with n_1, n_2, n_3, etc., individuals, then within these species two individuals can be selected in $n_1(n_1-1)$, $n_2(n_2-1)$ . . ., etc., different ways. The total number of different ways, therefore, in which different pairs of individuals of the same species can be selected from the population is half the sum of all these subsidiary totals, or in mathematical terms $\Sigma n \, (n - 1)/2$.

It follows that out of a total of $N \, (N - 1)/2$ possible different pairs $\Sigma n \, (n - 1)/2$ will have both individuals belonging to the same species. So on an average $N \, (N - 1)/\Sigma n \, (n - 1)$ pairs will have to be selected to get an even chance of obtaining one identical pair. This value can be called the Index of Diversity and is low when the diversity is small and high when the diversity is high. It can range, in this method of calculation, from "1" when all the individuals belong to one species (completely undiversified population) to "infinity" when every individual belongs to a different species (complete diversity).

This measure of diversity is dependent on the distribution of the individuals among the species as well as on the number of species present. For example, if we had a population made up of four species each with twenty-five individuals, giving a total of 100 individuals, the diversity would be $100 \times 99/4$ (25×24) $= 4.125$. On the other hand, had the species contained 90, 5, 3, and 2 individuals, with the same total of 100 individuals, the diversity would be only 1.19. Simpson's method is independent of any theory about the form of the frequency distribution of abundance, but is not entirely independent of the size of the sample (see p. 151). It also has the drawback of being very dependent on the numbers of the few more abundant species, and taking very little into account the rarer species, which, in most natural populations, make up a very considerable portion of the total species.

Prior to this suggestion two other direct approaches to the measurement of diversity had been made, the first by Fisher, Corbet and Williams in 1943, the second, for a different type of population, by Yule in 1944.

Our approach to the problem has already been described briefly (Chapter 3, p. 20) as arising out of the discovery that the logarithmic series showed a very close fit to the frequency distribution of species with different numbers of individuals in a random sample from an insect population.

The series, which can be expressed as

$$\alpha x, \ \alpha x^2/2, \ \alpha x^3/3, \ \alpha x^4/4 \ \ldots$$

contains two constants α and x, and it can be shown that x is dependent on the size of the sample, and gradually approaches unity as the sample size increases. The constant "α", on the other hand, is common to all samples from a single population, and so is a property of the population. It is higher when there is great diversity and low when the diversity is small, reaching "O" when all the individuals belong to one species. I called this the Index of Diversity, but prefer to extend the term to any other function with the same properties. In this particular form it is, of course, dependent on the existence of a logarithmic series distribution of abundance.

In 1944, Yule, studying the frequency distribution of the use of different nouns in samples from the writings of different authors, suggested as a "Characteristic" the value

$$10,000 \, \frac{M_2 - M_1}{(M_1)^2}$$

where M_1 and M_2 are the first and second moments of the distribution. This estimate is considered to be independent of the size of the sample. The multiplication by 10,000 was an arbitrary factor put in to avoid numbers with large numbers of decimal places. Yule's "characteristic" is high when the vocabulary is uniform, and low when there is great variety in the words used.

If, instead of this formula, we invert the moment's fraction and omit the multiplication factor, we get the value $\dfrac{(M_1)^2}{M_2 - M_1}$, which is high when the diversity is high and low when there is little variety. It cannot fall below unity. I have suggested calling this form Yule's Index of Diversity. It can be shown that it equals $N^2/\Sigma n\,(n-1)$.

In the case of the logarithmic series it should be noted that the number of groups with one unit, n_1, is equal to αx and as x approaches unity with large samples, n_1 approaches α. Thus with large samples from a population distributed in a log series the number of groups with one unit is an indication of the diversity of the population. Any single sample is, of course, subject to accidental error, but the closeness of the estimate can be seen in the case of Lepidoptera captured in a light trap at Rothamsted in the years 1933 to 1936. In these (see Table 66) the average diversity for the 4 single years was 38, and the average number of species per year with one individual was 39.

It has already been pointed out (p. 76) that in nearly all cases where samples of different sizes have been compared from the same population there is, over a considerable range, a more or less straight-line relation between the number of species represented and the logarithm of the size of the sample, this latter being measured either by the number of individuals, or (in the case of plants) by the area studied.

It will be seen from Fig. 126 (p. 311) that, given the same scale of measurement, the slope of the line representing this relation is a definite indication of

TABLE 66. *Lepidoptera captured in light traps at Rothamsted Experimental Station in different years, showing the number of individuals and species, the most abundant species, and the Diversity calculated on the basis of the logarithmic series*

Year	Indi-viduals	Species	Most abundant species No.	Most abundant species % of total	Index of Diversity "α"	Index of Diversity from moments	Species with 1 individ.
1933	3541	178	144	4.1	39.15	34.8	32
1934	3275	172	219	6.7	38.64	39.6	33
1935	6828	198	1799	26.3	38.19	11.3	37
1936	1977	154	275	13.9	39.05	24.0	54
4 years total	15,621	240	2347	15.0	40.24	25.5	34
4 years average	3902*	175	—	—	38.76	—	39

* Arithmetic mean: the geometric mean, 3537, is a better estimate.

diversity—the greater the diversity the steeper the line. Hence it follows that the rate of increase of species with increasing sample size is also a measure of diversity.

In the case of the log-series distribution the number of species in a random sample of N individuals is given by

$$S_1 = \alpha \log_e (1 + N/\alpha)$$

If, however, the sample size is sufficiently large for us to neglect the unit in relation to N/α (which it should always be to get a reliable sample), then the formula becomes

$$S_1 = \alpha \log_e (N/\alpha).$$

This is the situation in the range of the graph which approximates to a straight line. So if, in this range, two samples are taken, one p times the size of the other, then the increase in the number of species in the larger over the smaller sample is given by

$$S_p - S_1 = \alpha (\log_e pN/\alpha - \log_e N/\alpha) = \alpha \log_e p.$$

So by doubling the size of a sample at any level during this range the number of species added to the total is $\alpha \log_e 2 = 0.69\alpha$. If the sample size is increased by 10, the number of species added is 2.3α. If the sample size is multiplied by "e" (2.718), the increase in species would be (on an average) an exact measure of the index of diversity.

This holds over the range of the straight-line relation independently of any assumption about the form of the distribution.

Gleason in 1922 and 1925 approached the problem when he found a straight-line relation between number of species and the log of the area sampled, for plants (see also p. 67), and so the approximately constant increase in number of species for each constant multiplication of sample size.

Pidgeon and Ashby (1940) made an even closer approach when they proposed an empirical formula which can be written

$$S_p - S_1 = m \log_{10} p$$

to calculate the increase in number of species between one (S_1) and p (S_p) quadrats. Their m was a constant, and logs to the base 10 were used. It is clear, however, that their m is equivalent to $\alpha \log_e 10$ in our formula, and so equals 2.30α. This empirical relation does not hold for small samples.

Neither Gleason nor Pidgeon and Ashby realized the fundamental implications of their discovery, nor the wide applications of the concept of diversity as a measurable factor.

The Relation Between the Different Measurements of Diversity

Yule's Index $N^2/\Sigma n$ $(n - 1)$ can be written in terms of the first and second moments of the distribution in the form $(M_1)^2/(M_2 - M_1)$; the first moments being the total number of units in the sample.

For the log series the moments are $\alpha x/(1-x)$ and $\alpha x/(1-x)^2$ respectively from which it follows that

$$\text{Yule's Index} = \frac{(\alpha x)^2}{(1-x)^2} \Big/ \frac{\alpha x^2}{(1-x)} = \alpha.$$

Thus Yule's Index when applied to a logarithmic distribution gives a value identical with the α Index.

Simpson's Index when expressed as moments becomes

$$\frac{M_1{}^2 - M_1}{M_2 - M_1}$$

which in the case of the log series can be shown to be equal to

$$\alpha + 1 - 1/x.$$

The factor $1/x$ appearing in the formula shows that Simpson's estimate is not independent of the size of the sample. The value of x is related to the average number of units per group (individuals per species) in the sample. For example, if there are an average of only two units per group, the value of x is .71; if the average is 100, $x = .9984$. In the former case, usually found only in small samples, Simpson's Index would equal $\alpha - 0.4$; in the latter, it would be $\alpha - .001$.

Since Yule's measurement of diversity is $N^2/\Sigma n\,(n-1)$ and Simpson's is $N\,(N-1)/\Sigma n\,(n-1)$ it follows that Yule's value differs from Simpson's in the ratio of N to $(N-1)$; N being the total number of units in the sample. Thus the difference is greatest for small samples and almost negligible in large ones. It is 10 per cent higher in a sample of 10 units, but only 0.1 per cent higher in a sample of 1000 units. Since no estimate should be made from small samples the difference is, in practice, well below the error.

When biologists discovered empirically that the relation of species number to log individuals was (over a certain range) an approximately straight line, they did not comment on the fact that if this line were extended down to very small samples a point would be reached where a small but positive sample would have no species, or even less than none!

It follows from the log series that if the straight-line portion of the curve for any particular diversity is continued downwards to very small samples without the correcting factor of "1" in the formula

$$S = \alpha \log_e (1 + N/\alpha)$$

this extrapolation cuts the base line for zero species at a point equal to a population of α individuals. This is shown in Fig. 61 for $\alpha = 10$ and $\alpha = 50$.

Thus in a log series the size of this hypothetical sample with no species is still another measure of the diversity of the population sampled.

Although the index of diversity calculated from $M_1{}^2/(M_2 - M_1)$ is, for the logarithmic series, theoretically the same as "α", in practice, the two methods of estimation may give very different results. Consistent differences would indicate that the data did not strictly conform to the series, but there are also to be found differences in the variation between estimates from different samples from the same population which are, I think, due to the different weight put on the rarer and on the more abundant species by the two methods.

Theoretically, a log series, and others, envisage fractions of species with very large numbers of individuals per species. In practice, there must be a single most abundant species, and in different samples its value in number of individuals is very variable. This highest value has a very heavy weight in calculating moments, particularly with the second moment when it is squared;

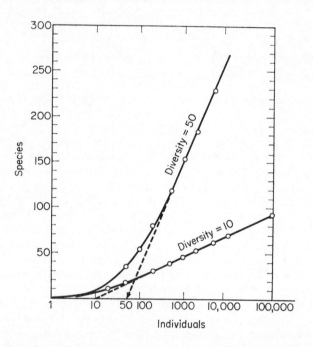

FIG. 61. The curve of the relation of species and log individuals on the basis of the logarithmic series, showing how the continuation of the straight-line portion of each curve cuts the line for one species at a number of individuals equal to the diversity. (See text for explanation.)

whereas in the calculating the value of α from the number of individuals and species, much more weight is given to the large numbers of the rarer species.

In captures of moths over 4 years in a light trap at Harpenden, in Hertfordshire, England, the number of individuals, the number of species, and the index of diversity from α and from the moments are shown in Table 66.

It will be seen that the values of the index of diversity for the 4 single years, calculated as α on the assumption of the log series, are remarkably consistent, and the slight increase in the 4-year total is exactly what would be expected in view of the greater diversity of any fauna over 4 years than in a single year.

The index from the moments is very variable indeed, and the value 11.3 in 1935 is particularly small. It is almost certainly due to the single most abundant species in that year having no fewer than 1799 individuals, which is over 26 per cent of the total catch. A calculation of the index omitting this one species gave a value of 28.7.

In the catches for 1936 and for the 4 years' total, the most abundant species were 13.9 per cent and 15 per cent of the total and in both the indexes are nearly equal to, but still well below, the values for 1933 and 1934, when no single species was so dominating.

If there is good reason, therefore, to assume the distribution to be of the log-series form, the estimate α for the diversity is much more consistent than that calculated directly from the moments.

SERIAL SAMPLING AND ITS EFFECT ON DIVERSITY

If an instantaneous sample can be taken from a population, one gets a reliable representation of the pattern at that moment, but unless the sample is large it is subject to considerable error. For example, it will be seen from Fig. 126 that a sample of fewer than 250 individuals from a log-series population gives a measure of diversity with an error of at least 10 per cent.

The ideal way of reducing the error is to increase the size of the sample without extending either the time taken or the range of material in the population sampled. This could be done by increasing the efficiency of the trapping method, for example of using a more efficient light trap, or a wider plankton net.

Under field conditions, however, it is often only possible to increase sample size by extending the period of time over which the sample is taken, or by taking simultaneous samples over a wider area. Each of these methods brings in greater information, but their effect on interpretation must be carefully watched.

One example of the effect of extending the time of sampling is shown in Table 67 and Fig. 62. This gives the number of moths caught in six traps in a small uniform area of woodland at Rothamsted from 30 June to 31 July 1949. The data given include the number of individuals and of species captured each night; and also the accumulated totals for each week and for the whole period of 32 days. In each case the changing diversity is also shown from the log series.

First it will be seen that the diversity of the single nights varies from 15 to 30 with an arithmetic mean value of 22.4 (Fig. 62 inset), with no evidence of any regular trend. At this time of the year (see p. 157), near the seasonal peak, no trend would be expected. For the whole month the 7375 individuals belonging to 197 species indicates a diversity of 37. This increase represents the greater natural diversity of the population when sampling is carried out over a complete month, during which earlier species have ceased to fly, and later ones have begun to appear.

For the separate weeks the diversities were 30, 28, 27, and 27; less variable

TABLE 67. *Captures of Lepidoptera for each of 32 days in six light traps at Rothamsted, with the accumulated weekly and months totals and the Diversities calculated on the basis of the logarithmic series*

Date	Each day			Accum. week		Accum. month		
	Ind.	Sp.	Div.	Ind.	Sp.	Ind.	Sp.	Div.
June 30	232	66	30	232	66	232	66	30
July 1	184	47	30	416	81	416	81	29
2	202	50	21	618	93	618	93	30
3	275	68	27	893	104	893	104	30
4	183	53	25	1076	110	1076	110	30.5
5	85	36	21	1161	112	1161	112	30.5
6	97	37	21	1258	115	1258	115	30.5
7	43	26	25	43	26	1301	115	30.5
8	111	38	21	154	46	1412	115	30
9	86	29	16	240	59	1498	118	30
10	134	49	29	374	71	1632	121	30
11	247	65	27	621	89	1879	129	31.5
12	377	67	24	998	103	2256	137	32.5
13	159	46	21	1157	106	2415	138	32
14	177	58	30	177	58	2592	144	33
15	187	50	23	364	76	2779	144	32
16	134	43	22	498	80	2913	146	32
17	95	33	18	593	83	3008	147	32
18	189	50	22	782	89	3197	149	32
19	93	37	22	875	91	3290	150	32
20	333	60	21	1208	104	3623	159	34
21	308	62	23	308	62	3931	162	34
22	534	75	23	842	90	4465	170	35
23	376	56	19	1218	98	4841	171	35
24	365	58	19.5	1583	107	5206	175	35
25	550	69	20.5	2133	119	5756	181	35.5
26	390	66	23	2523	127	6146	185	36
27	213	51	21	2736	135	6359	190	37
28	312	48	15.5	—	—	6671	191	37
29	146	38	16	—	—	6817	192	37
30	385	68	24	—	—	7202	197	37.5
31	173	43	18	—	—	7375	197	37

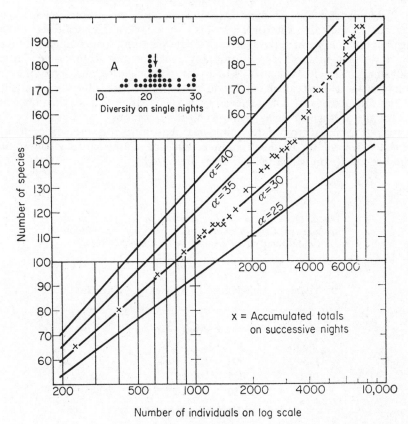

FIG. 62. Captures of Lepidoptera in six light traps at Rothamsted, Harpenden, during July 1949. Accumulated day-by-day total of individuals and species, showing the gradually increasing diversity as the period of sampling is extended. Inset A shows the frequency distribution of value of the diversity for each of the 32 nights individually.

than the single nights, but with an average 28.5 intermediate between the day and the month's values.

From Fig. 62 the steady increase in diversity as the accumulated total increases can be readily seen.

In this connection it is important to note that had the samples been taken at longer intervals over the same month, for example twice a week instead of every day, an apparently higher diversity would have been indicated, as the same range of population would have been sampled on fewer nights and hence with fewer individuals. It is essential that all comparisons should be made with identical sampling techniques.

Table 68 and Fig. 63 show a second example, covering a much longer period. In this, the number of moths, on a logarithmic scale, is plotted against the number of species for each single year, and for accumulated totals up to 8 years, in another moth-trapping experiment at Rothamsted. To get still smaller samples, values for $\frac{1}{8}$ of a year (from eight parallel series of every 8th night) were also calculated. It will be seen that there is the same tendency to increasing diversity with increasing time. The range is approximately from 39 for the average of single years, to 43 for a total of 7 or 8 years. As this would be expected from purely biological reasoning, it gives additional support to the validity of the technique, and the mathematical assumptions and arguments on which the estimates are based.

TABLE 68. *Catches in Lepidoptera in a light trap at Rothamsted Experimental Station, showing the relation of the diversity to the length of the trapping period*

Year	Individuals	Species	Diversity
One-eighth of a year, average			
1933	440	89.6	34.0
1934	409	86.8	33.7
1935	855	107.1	32.2
1936	250	67.7	30.9
Whole year			
1933	3454	173	38.4
1934	3276	168	37.4
1935	6530	191	36.8
1936	1961	154	39.2
1946	3124	156	35.1
1947	3786	184	40.2
1948	3771	187	41.1
1949	6107	191	37.5
4 years			
1933–36	15,221	234	39.7
1946–49	16,972	254	42.9
All 8 years	32,853	285	43.8

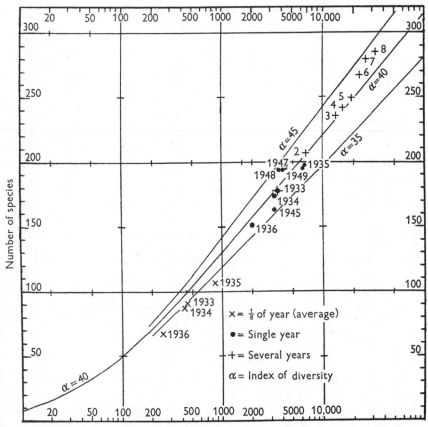

FIG. 63. Accumulated totals of individuals and species in captures of Lepidoptera in a light trap at Rothamsted over periods of from 1/8 to 8 years, showing slowly increasing diversity with increasing period of time over which the trapping was continued. (From *J. Anim. Ecol.* (1953), **22**, 22.)

THE ANNUAL CYCLE OF DIVERSITY IN LEPIDOPTERA IN BRITAIN

Lepidoptera were captured in light traps at Rothamsted, about 25 miles north of London, every night for two periods of 4 years, 1933–36 and 1946–49. Table 69 shows, for each month, the diversity calculated on the assumption of a logarithmic series distribution. The seasonal changes are shown graphically in Fig. 64. The numbers captured during the winter months were too low to give reliable estimates, but it can be assumed (see below) that the diversity was very low at this season.

It will be seen that the population diversity reaches its peak in July with

TABLE 69. *Seasonal changes Index of Diversity in captures of Lepidoptera in 8 years at Rothamsted (see Fig. 64)*

	Apr.	May	June	July	Aug.	Sep.	Oct.
1933	3.5	11	22	25	16	8.8	7.0
1934	2.9	12	18	28	14.5	7.5	4.0
1935	5.1	11	16	25	10.5	8.0	5.0
1936	3.0	12	12.5	22.5	12.5	8.5	3.7
4-year mean	3.8	11.5	17.1	25.1	13.4	8.2	4.9
1946	—	8	12	24	14	7.5	6.0
1947	2.2	12	23	28	17	6.5	4.5
1948	3.2	15	17.5	30.5	21	5.2	6.0
1949	3.5	13.5	21.5	26	12.5	8.0	5.8
4-year mean	3.0	12.1	18.5	27.1	16.1	6.8	5.6
8-year mean	3.3	11.8	17.8	26.1	14.8	7.5	5.3

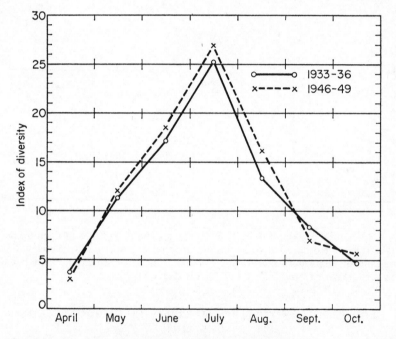

FIG. 64. Seasonal change in specific diversity of Lepidoptera at Rothamsted from captures in a light trap during two periods of 4 years each, 1933–6 and 1946–9. The diversities are calculated on the assumption of the log series.

an average of 26, as compared with 3.2 in April and 5.3 in October. These values indicate that if uniform samples of 1000 insects had been taken in each month, the number of species represented would have been about ninety-five in July, as compared with nineteen in April and thirty in October.

The variation from year to year in the same month appears to be greater in the early summer than in the autumn. It is particularly large in June, with a range from twelve to twenty-three, and low in April, September and October.

In confirmation of these Rothamsted results, information is available for catches of Lepidoptera at Woking, about 17 miles south-west of London, made by Best (1950). He had two light traps, the first of which was at a height of 4½ ft above the ground and was in use for 30 months from May 1947 to November 1949, during which 12,009 moths of 342 species were captured, indicating a diversity of 65.

The second trap, at a height of 36 ft in the same garden, was run simultaneously with the lower trap, but only for the 21 months from March 1948 to November 1949. During this period 10,606 individuals belonging to 336 species were captured, giving a diversity of approximately 66. It should be noted, however, that the first trap period included 3 summers and 2 winters, and the second trap 2 summers and 1 winter, so that both give average values above the normal for a complete year: the upper trap to a greater extent than the lower.

Table 70 shows the results for the 21 months during which both traps were

TABLE 70. *Seasonal changes in diversity of Lepidoptera during 2 years at Woking, Surrey, in two traps; one at 4½ ft and the other at 36 ft above the ground. Values for catches in Harpenden for corresponding months are given for comparison*

| | Lower trap | | | Upper trap | | | Rotham-sted |
	Individ.	Sp.	Div.	Individ.	Sp.	Div.	Div.
1948							
March	518	17	3.3	105	10	2.5	
April	121	26	10	160	26	9	3.2
May	205	58	28	504	76	25	15
June	506	80	28	1034	95	26	17.5
July	838	111	34	1737	125	30	30.5
August	392	73	27	828	92	26	21
September	271	37	11	461	47	13.5	5.2
October	67	19	10	167	30	11	6
November	185	9	5	144	16	4.5	
December	79	3	1	12	3	2	
1949							
January	55	6	2	2	2	—	
February	51	10	4	21	4	1.5	
March	234	14	3.5	160	11	2.5	
April	83	24	12	58	18	9	3.5
May	135	42	20	271	45	15	13.6
June	620	78	24	1187	99	26	21.5
July	636	104	35	1334	125	34	26
August	422	67	22	1436	81	19	12.5
September	272	35	11	789	47	11.5	8.0
October	100	13	4.5	169	19	5.5	5.8
November	88	8	2	32	6	2	

working, and Fig. 65 shows the same graphically. The diversity for the corresponding months at Rothamsted has been added when available. The Woking monthly diversities are all slightly greater than the corresponding Rothamsted figures, which is very possibly due to a greater diversity of local vegetation in the former locality.

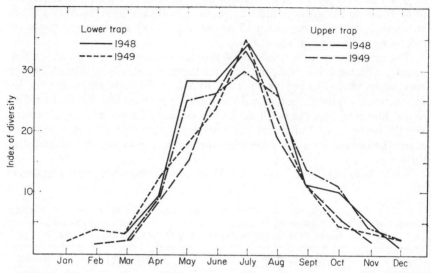

FIG. 65. Seasonal changes in specific diversity of Lepidoptera captured in two light traps at Woking, England, during 1948 and 1949, one at 4½ ft and the other at 36 ft above the ground (Best, 1950).

DIVERSITY OF LEPIDOPTERA IN HOLLAND

To give a further example, this time from outside the British Isles, Table 71 shows the captures of Macro-Lepidoptera in a light trap in the Netherlands made by Houtman (1961) over 3 successive years. The seasonal changes in

TABLE 71. *Seasonal changes of Diversity in Lepidoptera as illustrated by 3 years' captures in a light trap in the Netherlands by Houtman*

	1954			1955			1956			Diversity arith.
	Ind.	Spec.	Div.	Ind.	Spec.	Div.	Ind.	Spec.	Div.	mean
March	—	—	—	—	—	—	30	6	1.8	
April	—	—	—	138	11	2.7	58	3	1.1	
May	758	37	8.5	205	24	7.5	438	38	10	9
June	2828	75	14	2015	57	11	3181	65	11	12
July	2197	66	13	2419	102	23	6521	112	19	19
August	7890	129	22	6157	109	19	2860	76	14	18
Sep.	2585	62	11	1524	29	5	2060	54	11	9
Oct.	340	17	4	140	15	4	213	14	3.5	4
Nov.	14	2	1	32	8	4	13	5	4	3

specific diversity are almost exactly similar to those given above, with the peak in 2 years in July and in 1 year in August.

SPECIFIC DIVERSITY OF LEAFHOPPERS IN FINLAND

Kontkanen (1950) published the results of a field study of leafhoppers (Homoptera-Auchenorrhyncha) in North Karelia, Finland, in which he gives details of the results of regular sampling by means of sweep-nets in 3 summers. Some of his data are shown in Table 72, from which the diversities have been calculated, and the seasonal changes are shown diagrammatically in Fig. 66.

TABLE 72. *Seasonal changes in specific diversity in leafhoppers captured in 3 years in Finland*

Sample No.	Date	Individ.	Sp.	Diversity
	1942			
1	June 17–18	282	16	3.6
2	June 28–29	631	20	3.9
3	July 18–20	2128	38	6.6
4	July 30–Aug. 2	2912	45	7.6
5	Aug. 13–15	2019	48	8.8
6	Aug. 25–26	1350	51	10.1
7	Sep. 5–6	665	41	9.5
	1947			
1	June 7–13	2146	16	2.4
2	June 24–27	2870	30	4.6
3	July 12–17	3848	54	8.9
4	Aug. 1–3	3642	55	9.2
5	Aug. 15	1913	43	7.8
6	Aug. 28	348	37	10.4
7	Sep. 14–15	255	34	10.5
	1948			
1	June 19–20	1355	32	5.9
2	June 26–27	3044	45	7.5
3	July 13	4288	56	9.1
4	July 25	2927	56	9.9
5	Aug. 11	2526	52	9.3
6	Aug. 27	585	41	9.8
7	Sep. 13	255	29	8.3

It will be seen that there is a steady rise in specific diversity from an average of about 4 in June to about 9 at the end of July, and to 10 at the end of August. There is some indication of a small fall in September, but the population was then falling rapidly, thus increasing the error. The peak of abundance of individuals was at the end of July, so the diversity continues to rise after the population has begun to fall.

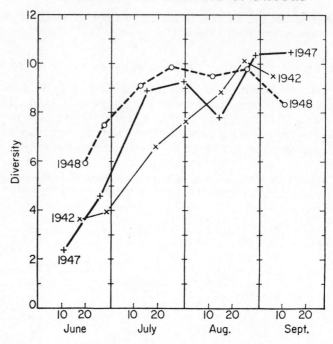

FIG. 66. Seasonal changes in specific diversity of leafhoppers in Finland, from June to September in 3 years (Kontkanen, 1950).

TABLE 73. *Relation between numbers of individuals, species, and genera in samples of leafhoppers from thirty different biotypes in Finland*

Samp.	Ind.	Sp.	Gen.	Samp.	Ind.	Sp.	Gen.	Samp.	Ind.	Sp.	Gen.
1	107	6	5	11	34	10	10	21	68	15	15
2	64	8	8	12	87	10	10	22	346	25	22
3	49	7	7	13	82	11	11	23	1292	20	17
4	339	11	10	14	235	16	14	24	434	16	14
5	82	8	8	15	158	16	15	25	822	21	17
6	56	11	11	16	372	18	17	26	335	21	19
7	61	7	7	17	156·	15	13	27	542	15	13
8	292	9	9	18	461	19	16	28	498	19	16
9	477	13	12	19	434	22	18	29	329	14	12
10	435	9	8	20	535	18	15	30	644	22	17

Total in all 30 samples: 9826 individuals, 87 species and 52 genera

In Table 73 and Fig. 67 there is shown the relation between number of species and number of individuals in samples taken from thirty different biotypes. In the single samples the diversity ranges from 1.5 to 6, approximately, the error in those samples with less than 100 individuals being greater. If all the samples are added together we get eighty-seven species with 9826 individuals, which has a diversity of 13.4 as shown in the upper corner of the figure. Thus the index of diversity once more reflects the diversity of the environment.

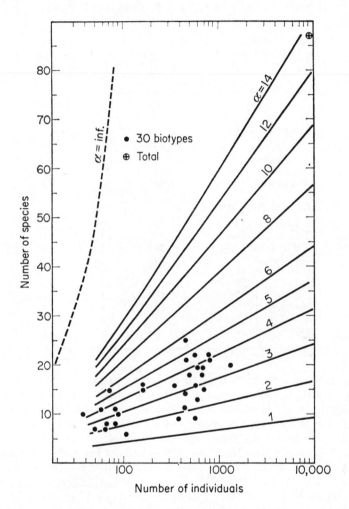

FIG. 67. Specific diversity of leafhoppers in Finland in thirty different biotypes, and in the total of all samples (Kontkanen, 1950).

CADDIS FLIES CAPTURED IN A LIGHT TRAP AT READING

In 1960, Crighton gave particulars of catches of Trichoptera (excluding the Hydroptilidae) in a light trap at Reading, about 35 miles west of London, in 3 successive years 1953–5. In the 1st year the trap was in use every 3rd night from 26 May to 27 July, and from 21 August to 13 December, on a total of 49 nights. In 1954 it was in operation every 3rd night from 18 April to 24 November, with a total of 78 nights. In 1955 every 3rd night from 21 April to 6 December, again on 78 nights. Except for the later start in 1953 the trapping period overlapped the beginning and end of the insect broods.

TABLE 74. *Number of Caddis flies* (Trichoptera) *captured in light traps near Reading in certain years, and on certain nights of high catch, together with the Diversities calculated on a log-series distribution*

	Individ.	Sp.	Diversity
Complete years			
1953	2635	43	7.1
1954	5934	46	6.9
1955	8419	47	6.8
Total 3 years	16,988	59	7.6
Nights with large catches			
15–16 Sep. 1953	437	10	1.8
12–13 Oct. 1953	363	10	1.9
9–10 Oct. 1954	363	10	1.9
30–31 Sep. 1954	243	7	1.5
15–16 Oct. 1954	237	9	1.8
20–21 Sep. 1953	232	10	2.0
22–23 Sep. 1953	174	9	2.0
21–22 Sep. 1955	162	8	1.8
16–17 Sep. 1953	130	9	2.2
18–19 Oct. 1954	108	8	2.0

A summary of his results is given in Table 74 from which it will be seen that the diversity in each of the 3 years is very similar. The 1st year, at 7.1, is a little higher, and associated with a much smaller number of individuals. Although this difference is probably not significant, it is interesting to note that a small increase in diversity might well be expected, since the season was covered by fewer samples. Thus species that had not reached their peak when trapping stopped at the end of July, or had passed their peak when the trapping was restarted at the end of August would be represented, but by fewer individuals. So even this small difference might be expected, but as an artificial one, not due to biological events.

The increase in the diversity to 7.6 for the 3 years together is to be expected from the greater variety of environment, and particularly the climatic conditions, in a series of years.

Dr Crighton has also given to me particulars of catches on a number of single nights with exceptionally high activity, which are also summarized in the same table. Once again the calculated diversities are remarkably consistent, varying from 1.5 to 2.2 with an average value of 1.9.

DIVERSITY OF MOSQUITO POPULATIONS IN CITIES IN IOWA,
U.S.A.

In 1942, Rowe gave the numbers of mosquitoes caught in light traps in ten cities in the State of Iowa. His information has been summarized in the first two columns of Table 75 and shown graphically in Fig. 68, where the log of the number of individuals is plotted against the actual number of species. In the third column of the table is shown the index of diversity of each sample and of the total for all cities. These are subject to a standard error of from 7.5 to 10 per cent, the smaller error being in the larger samples. It will be seen that for the single cities the values are grouped round an average value of 2 and none differs significantly from that value.

TABLE 75. *Number of individuals, number of species, and diversity in samples of mosquitoes from cities in Iowa, U.S.A.*

	No. of individ.	No. of sp.	Index of diversity
Ruthven	20,239	18	1.95
Res Moines	17,077	20	2.24
Davenport	15,280	18	2.02
Ames	12,504	16	1.81
Muscatine	6426	16	1.98
Dubuque	6128	14	1.71
Lansing	5564	16	2.03
Bluffs	1756	13	1.90
Sioux City	661	12	2.08
Burlington	595	12	2.13
All cities	86,230	28	2.70

To Fig. 68 has been added lines showing the theoretical relationship between species and individuals for log series diversities of 1, 2 and 3, and the closeness of fit of the different samples to the value $\alpha = 2$ is apparent. From this we can see that the extent to which the individuals are spread among species is more or less the same in all the cities; the small number of species caught in some, and the larger number in others, is almost exactly accounted for by the varying number of individuals captured.

When, however, all the ten catches are added together we get 86,230 individuals of twenty-eight species showing a diversity of 2.7 which is significantly different from the average of 2 for the separate cities. From this we can infer that although the cities are similar in their diversification, they have not identical populations. The area considered as a whole has a greater variety of environment than any one of its parts.

Although the lines for the different values of α in the figure have been calculated from the log series, it is important to note that in this straight-line portion as represented by such large samples (and particularly the very large

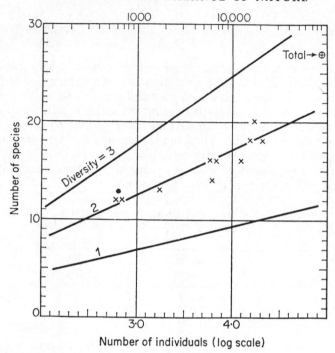

FIG. 68. Relation between the number of individuals (on log scale) and number of species, in samples of mosquitoes from ten cities in Iowa, U.S.A., showing their relation to diversity as estimated from the log series.

value of N/α) they can be taken to represent conditions in any distribution which presents a straight-line relation in this range.

It is interesting to note that if two samples contain approximately the same number of individuals, then the number of species in each can give a general indication of diversity. For example, in the present series the two towns, Muscatine and Dubuque, have given nearly the same-sized samples, 6426 and 6128, but the difference in the number of species (16 and 14), though small, is considerable for such large samples. Thus one would expect a slightly smaller diversity in Dubuque, even though in this case it is not significant.

It is when we wish to compare samples of different sizes that the method of inspection breaks down. For example, does the catch of 15,280 mosquitoes of eighteen species from Davenport indicate a higher or a lower diversity than the 661 individuals of twelve species at Sioux City? The mathematical answer is that they represent almost identical diversities of 2.02 and 2.08. In other words had a sample of only 661 mosquitoes been taken from Davenport, one would have expected it to contain twelve species as actually caught at Sioux City.

DROSOPHILIDAE IN THE UNITED STATES AND MEXICO

Patterson, in a series of tables (1943), gave captures of Drosophilidae in a number of States of the U.S.A., and in a number of localities in Mexico. The information about the number of individuals and the number of species in each area is summarized in Table 76 together with the calculated diversity on the assumption of the log series.

It will be seen that the diversities range from 1.3 to 4.7 for single States in the U.S., and from 2 to 8 for single localities in Mexico. When several States or localities are combined there is the expected rise in diversity due to a greater variety of environment. There is, as shown in Fig. 69, a general

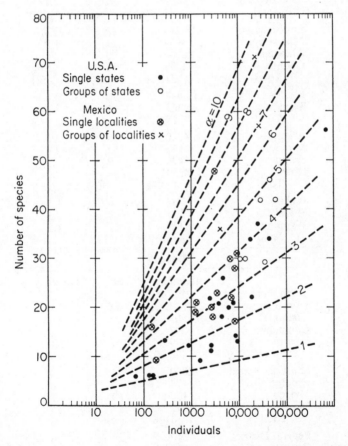

FIG. 69. The relation between the numbers of individuals (on a log scale) and the number of species in samples of Drosophilidae captured in different areas in U.S.A. and Mexico; and their relation to the Index of Diversity as calculated from the logarithmic series (Patterson, 1943).

TABLE 76. *Individuals, species, and diversity of samples of Drosophilidae from different areas in U.S.A. and Mexico*

UNITED STATES			
	Individ.	Sp.	Diversity
Texas	674,675	56	4.7
Oklahoma	2255	22	3.4
Colorado	8718	14	1.7
Utah	7406	21	2.7
Nevada	124	6	1.3
California	9584	13	1.5
Total	28,087	42	4.9
Arizona	26,747	37	4.2
New Mexico	16,492	34	4.1
Total	43,239	46	5.0
Louisiana	18,289	22	2.5
Mississippi	3962	21	2.9
Alabama	6180	20	2.8
Georgia	2685	13	1.8
Total	31,116	29	3.1
Florida	13,447	30	3.6
Tennessee	45,767	34	3.6
Total	59,215	42	4.4
Arkansas	4320	18	2.5
Missouri	946	12	2.0
Kentucky	1420	9	1.3
Ohio	4389	26	3.6
Total	11,075	30	3.8
Idaho	67	6	1.5
S. Dakota	134	6	1.3
Wyoming	2390	11	1.5
Total	2591	12	1.6

MEXICO			
	Individ.	Sp.	Diversity
Tamaulipas	2193	20	3.2
Nuevo Leon	7633	22	2.7
Coahuila	2803	18	2.6
Chihuahua	6297	30	4.1
Sonora	7979	17	2.1
Total	26,905	57	6.7
San Luiz Potosi	2880	23	3.4
Hidalgo	174	9	2.0
Mexico	130	16	4.7
Morelos	1242	21	3.5
Total	4426	36	5.5
Federal District	8640	28	3.6
Guerro	9530	31	4.0
Michoa Chan	3338	48	8.0
Jalisco	1210	19	3.2
Total	22,718	71	9.2

tendency for the diversity to increase with the size of the catch; this may be due in part to the larger catches being more diversified either by time (season) or by area. Texas gives the highest diversity for any single State in the U.S., but in view of the very large catch it seems likely that a greater variety of locations was tested. The Mexican localities have a higher average diversity, and the three groups and one locality (Michoa Chan) give higher diversities than any collection in the United States. This undoubtedly reflects the general increase in specific diversity in warmer climates.

MIGRANT AND NON-MIGRANT LEPIDOPTERA IN ISRAEL

Lane and Rothschild (1961) caught Lepidoptera in a light trap at Herzlia, on the Mediterranean coast of Israel, from 22 March to 10 April. They trapped a total of just over 6300 moths belonging to fifty-seven species, of which full details are given in their paper. As the trapping was done in the spring, when one might expect a northerly movement of migrant insects, they sorted the species according to their known habits; this resulted in thirty species being considered as probable migrants, twenty-six as probable residents and one of doubtful status.

An analysis of their data according to the frequency distribution of species with different numbers of individuals is shown in Table 77 and in Fig. 70.

It will be seen that there is the familiar hollow curve of abundance, but also that there is a great difference between the general pattern of the migrants and of the resident species. The thirty migrant species are represented

TABLE 77. *Frequency distribution of migrant and non-migrant Lepidoptera, caught in a light trap in the spring on the coast of Israel*

Individ. per sp.	No. of sp. migrant	No. of sp. Non-migrant	Individ. per sp.	No. of sp. migrant	No. of sp. Non-migrant
1	4 (4)	12 (12)	38	1 (4)	— (2)
2	3	3	47	1	—
3	4	—	57	1	—
4	1 (8)	1 (4)	58	1	—
5	—	1	121	1 (4)	— (0)
6	—	1	142	1	—
7	—	1	207	—	1
8	1	3	242	1 (2)	— (1)
10	— (1)	1 (7)	370	1	—
14	—	1	589	1 (2)	—
15	1	—	1832	1	—
19	—	1	2398	1 (2)	—
22	1	—	Total species	27*	26
29	1	—	Individuals	5944	315
			Diversity (approx.)	3.7	7.0

* Also three migrant species with "several" individuals.

by approximately 3000 individuals, with an arithmetic mean of 200 per species. The twenty-six resident species are represented by only 315 moths with a mean of 12.1. Of this small total 207 belong to a single species *Siderides scirpi* Dup., which belongs to a genus containing a number of well-known migrants; it is possible that its assumed status as a non-migrant is only an expression of our ignorance. It is the only species among the residents of which more than twenty individuals were captured, whereas thirteen of the thirty migrant species had more than this number.

A calculation of the diversities gives a value of 3.2 for the migrants and 7 for the resident population, from which we get the expected number of species with one individual (n_1) to be 3.2 for the migrants (observed, 4) and 6.6 (observed, 12) for the residents. If, however, the uncertain *S. scirpi* is excluded from the residents, we get an estimated diversity of 10.3, and an n_1 of 9.4, which is considerably closer to the observed number.

So it appears that in this area, and at this time of the year, the possibly migrant population of Lepidoptera not only equals the resident population in species, but greatly outnumbers them in the number of individuals.

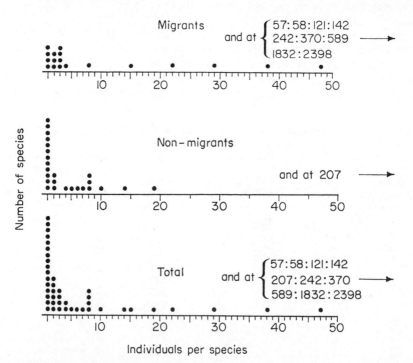

Fig. 70. Frequency distribution of species with different numbers of individuals of Lepidoptera taken in a light trap on the coast of Israel during March and April 1961. Also separate results for the species which are considered to be probable migrants or resident, showing the difference in relative abundance between the two (Lane and Rothschild, 1961).

On the other hand, the migrant population is distinctly less diverse, having fewer species in relation to the number of individuals.

DIVERSITY OF SMALL MAMMALS IN QUEENSLAND AND MALAYA

Harrison (1960) has made a comparison of the numbers of individuals and species of small mammals, collected mostly by trapping, in Queensland and in Malaya. The methods of trapping were similar in both countries, so that any differences probably indicate real biological variation. A summary of his results is given in Table 78 and diagrammatically in Fig. 71. The calculations of diversities with their errors are his.

TABLE 78. *Diversity of small mammals in different types of country in Queensland (Q) and Malaya (M)*

Habitat	Area and country	Individ.	Sp.	Diversity
Rain forest	Ulu Langat (M)	3093	31	4.9 ± 0.3
	Ampang Expt. (M)	183	13	3.2 ± 0.5
	S.B. Forest Expt. (M)	196	13	3.1 ± 0.5
	Innisfail Dist. (Q)	142	7	1.5 ± 0.3
	"D" Expt. (Q)	123	6	1.3 ± 0.3
Secondary forest	Sungel Menyala (M)	117	10	2.6 ± 0.6
	Guillemard Dist. (M)	68	7	1.9 ± 0.5
	Innisfail Dist. (Q)	42	6	1.9 ± 0.6
	"H" Expt. (Q)	165	9	2.0 ± 0.4
Grassland	S.B. Pylon Expt. (M)	233	5	0.9 ± 0.2
	S.B. Stream Expt. (M)	186	5	0.95 ± 0.2
	Innisfail Dist. (Q)	22	5	1.3 ± 0.3*
Sugar cane	"S" Expt. (Q)	163	7	1.5 ± 0.3
Town house	Kuala Lumpur (M)	10,610	5	0.5 ± 0.08
	Innisfail (Q)	24	2	0.5 ± 0.2

* Thus in original. I estimate diversity about 2.1.

Taking first the rain forest association, the diversities in the three areas in Malaya were 4.9, 3.2 and 3.1, while two localities in Queensland were only 1.5 and 1.3. In secondary forest the differences were less and probably not significant: two in Malaya being 2.6 and 1.9, two in Queensland 1.9 and 2.0. Grassland trapping gave a small difference in the reverse direction, but Harrison comments that the type of grassland in Malaya, dominated by Lalang Grass (Imperator), is very different from the Queensland association which is much more varied.

The results from trapping small mammals in houses is remarkable in that one sample of over 10,000 individuals in Kuala Lumpur and one of only twenty-four individuals from Queensland gave almost identical diversities, and, as would be expected from the more restricted environment, lower than the outdoor samples.

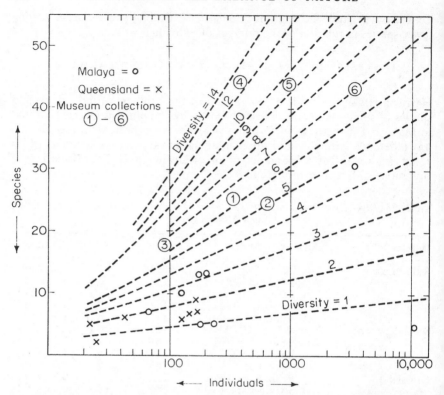

Fig. 71. Relation between number of individuals, number of species, and the diversity in collections of small mammals from different areas in Malaya and in Queensland (Harrison, 1960).

In a second set of data, shown in Table 79, Harrison gives details of some general collections from the same part of the world and made for museums. He comments that though these are of interest, they are not as reliable as those already discussed, as collectors for museums are apt to concentrate on the rarer species. The highest diversity in these collections is 12.9 for a collection from the mainland of Borneo; next, at 9.5 and 7, come two collections from Malaya; while two collections from Queensland bring up the rear with "open forest" 6.4 and "rain forest" 4.9.

Harrison comments on his results as follows: "There seems therefore to be good evidence that the tropical rain forest of North Queensland is deficient in a number of species of mammals, certainly as compared with the botanically very similar Malaysian rain forests. Such a lack of diversity, if true of other animals, might help to explain such phenomena as the outbreaks of insects in the rain forests of New South Wales recorded . . . by Brereton (1957). Such outbreaks are not usual in Malaysian rain forest, but the reduced flora of the temperate forest may well be accompanied by an even more severely reduced fauna; conditions suitable for violent oscillations in animal numbers. A comparison of the fauna of the Australian rain forest inside and outside of the tropics would seem desirable."

TABLE 79. *Diversity of small mammals in Queensland, Malaya and Borneo from published lists from museums. Possibly biased in favour of rare species*

Habitat	Country	Individ.	Sp.	Diversity
Open forest	(1) Cape York Peninsula	320	25	6.4 ± 0.7
Rain forest	(2) Cape York Peninsula	646	24	4.9 ± 0.5
	(3) Natuna Islands (W. of Borneo)	98	18	6.5 ± 1.1
	(4) Borneo, mainland	361	44	12.9 ± 1.1
	(5) Ulu Gombak, Malaya	987	44	9.5 ± 0.7
	(6) Ulu Langat, Malaya	3189	43	7.0 ± 0.5

DIVERSITY IN BENTHIC FAUNA IN RELATION TO DEPTH

In 1951 I dealt briefly (Williams, 1951b, p. 137) with some figures given by Sverdrup (Sverdrup et al., 1942, p. 806) on the number of individuals and species found at a number of stations at various depths during the voyages of H.M.S. *Challenger*. His summary included, for example, the statement that at twenty-five stations over 4500 m deep there was an average catch of twenty-four individuals and 9.4 species per station. Later, on referring to the original data in Murray (1895, p. 1430), I found that his figures for this depth were a total of 600 individuals belonging to 340 species in all the twenty-four stations together. It is permissible to divide the number of individuals by the number of stations to get an average value, because no individual occurs in more than one station; but to divide the number of species by the number of stations gives a meaningless and misleading figure. Thus my calculations based on Sverdrup's table were invalidated.

The correct figures, as given by Murray, are shown in Table 80 and Fig. 72. It is unfortunately not possible to get from his table the actual number of

TABLE 80. *Specific diversity in Benthic animals from records of H.M.S. Challenger*

Depth in metres	No. of stations	No. of specimens	No. of species	Specific diversity
180–900	40	6000	2050	1100
900–1800	23	2000	710	398
1800–2700	25	2000	600	299
2700–3600	32	1250	500	313
3600–4500	32	820	340	218
Over 4500	25	600	235	145

species at each station, from which an average per station could be calculated; but it is possible to find the diversity at each depth for all the samples at that depth, and these are shown on the table. It will be seen that there is an almost continuous decrease in specific diversity with increasing depth, the rate of decrease being particularly rapid between 500 and 1200 m. Unfortunately no total of individuals was given for the stations less than 180 m, though the total species was given. The very high values of the diversity are due to the great range of the animal kingdom covered, and the great geographical range of the stations. The problem of generic diversity is discussed on p. 180.

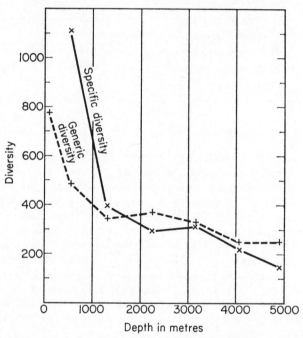

FIG. 72. Changes in specific and generic diversity of sea-floor animals at different depths, from data in the *Challenger* Reports (Murray, 1895).

DIVERSITY FROM PLANT QUADRATS WHEN THE NUMBER OF INDIVIDUALS IS UNKNOWN

Blackman (1935) gives the average number of species of plants found on quadrats of different sizes on a grassland formation at Bracknell, Berkshire, England, in 1932, as shown in Table 81.

The quadrats are small, but against this can be put an increased accuracy in assessing the number of species. A series of doublings in size from 2 to 128 in² gives increases in the number of species as shown in the third column. The increases are irregular, but there is some evidence that the earlier doublings produce less effect, which would be expected on theoretical grounds.

Assuming an approximately straight-line relation between the log of sample size and the number of species, the index of diversity can be calculated from the formula $S = \alpha \log_e (N/\alpha)$, if we consider that the actual number of individuals is proportional to the area sampled.

If we take the average increase in number of species per doubling in the above data as 2.05 we get the diversity

$$2.05/\log_e 2 = 2.96.$$

TABLE 81. *Number of species of plants in areas of different sizes, in plant associations in England*

Size of quadrat in sq in	No. of sp.	Increase of sp. on doubling
2	5.9	
4	7.8	1.9
8	9.5	1.7
16	11.1	1.6
32	13.6	2.5
64	16.1	2.5
128	18.2	2.1
	Average increase	2.05

If we plot, for a diversity of 3, log individuals against number of species, we get the line as shown in Fig. 73. If we further make the assumption that each square inch of the area sampled by Blackman contained on an average

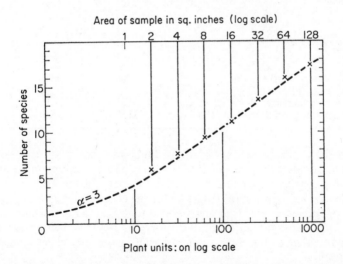

FIG. 73. Number of species of flowering plants on small quadrats of different sizes, recorded in Berkshire, England, by Blackman (1935), showing the relation between increasing area (on log scale) and increasing number of species.

eight individuals, his observed values fit almost exactly on this line of diversity 3. As he found an average of 5.9 species of plants on each 2-in² plot the suggestion of eight plants per square inch is not unreasonable.

Thus Blackman's data can be quantitatively analysed to suggest that the association sampled had a diversity of about 3, and contained about eight plants per square inch, or 1150 per square foot.

Area and Number of Species in an Upland Forest in Mauritius

In 1941, Vaughan and Wiehe gave details of the number of species of forest trees in a series of quadrats of increasing size in three associations in an upland climax forest in the island of Mauritius. These numbers are shown in Table 82, together with the total number of trees actually counted on 1000 m² in each of the associations.

TABLE 82. *Relation between area and number of species in three associations of forest plants in Mauritius*

| Area in sq m | No. of species | | |
	A Heath Assoc.	B Sideroxylon Assoc.	C Climax Plot 1
2	—	6	3
4	4	9	5
10	6	16	10
20	10	23	15
40	12	34	22
80	16	46	28
120	18	53	32
200	19	61	38
520	28	73	56
1000	30	79	70
Observed plants on 1000 sq m	2206	3838	1785
Diversity from sp. on 1000 sq m	5	14	14.5
Diversity from increase 40–400 sq m	5	15	11
Estimated plants per 1000 m from this latter	2130	3550	1680

From the observed number of individuals and species on this area it is possible to calculate the index of diversity for each association, on the assumption of the logarithmic series. This is found to be 5 for the heath association (A), 14 for the Sideroxylon (B), and 14.5 for the climax forest, plot 1. The total number of species on the three areas added together was 114, which, with the total of 7829 trees, gives an overall diversity of 19. This, being considerably larger than either of the three single associations, is a confirmation of their differentiation.

An alternative approach is to calculate the diversity from the average rate of increase in species between 20 and 200 m². This (see table) gives an identical value of 5 for the heath association, a slightly higher diversity, 15, for the

Sideroxylon, and a distinctly lower value of 11 for the climax forest. From these diversities and the number of species on 200 m² the number of individuals per square metre was estimated as shown in the table. The results for all three are slightly below the observed values.

Figure 74 shows the observed number of species in the three associations in relation to increasing sample size, as given by Vaughan and Wiehe, together with lines, calculated from the log series, to fit the estimated diversities and the number of individuals per unit area. The large difference between the diversity of plot c, the upper climax forest, as calculated from the observed number of individuals, and that from the rate of increase, can be partly accounted for by the large increase in the number of species between 200 and 1000 m². This may possibly be due to an increase in the diversity of the environment as the area increased. Apart from this there is a close fit between the theoretical and the observed values.

In the same paper, Vaughan and Wiehe give the number of species on ten different plots of 1000 m² each (50 × 20 m) in the climax forest association, and also the number of individuals. In this, however, only trees with a diameter of 10 cm or more at breast level (about 1 m) were counted. This results

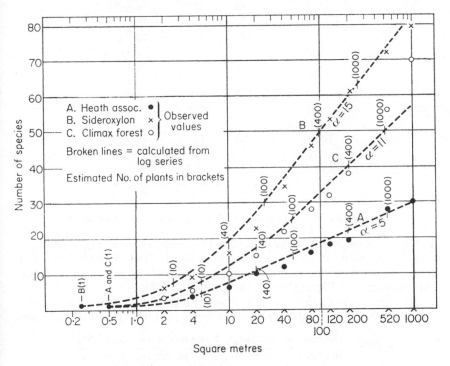

Square metres

FIG. 74. Relation between area (on log scale), number of species of forest trees, and the index of diversity in different plant associations in Mauritius. Estimated number of trees given in brackets.

in a very great reduction in the number of plants counted, but only a slight reduction in the number of species. The average for the ten plots of 1000 m² is 171 individuals of 28.5 species (diversity 9.7), as compared with 2206 individuals of thirty species (diversity 5) in plot 1 of the same series dealt with more fully above, in which all the smaller plants were included.

Table 83 shows the variation from plot to plot in the diversity, and in Fig. 75 the results are shown diagrammatically, together with the theoretical lines for diversities of 8, 9, 10 and 12. The very close fit of the diversity of the average and the total of all ten plots to the value 10 indicates a considerable uniformity of vegetation over the whole area.

TABLE 83. *Number of individuals and species and the diversity of forest trees in different plots of a climax association in Mauritius*

Plot No.	Individ.	Sp.	Diversity
1	126	30	12.0
2	163	24	8.0
3	163	27	9.5
4	173	26	8.5
5	153	29	10.5
6	166	26	8.5
7	199	32	10.5
8	186	29	9.5
9	167	31	11.0
10	214	30	9.5
Total	1710	52	10.0
Average	171	28.5	9.7

JACCARD'S COEFFICIENT OF FLORAL COMMUNITY

In a series of papers published between 1902 and 1941, Jaccard, largely as a result of the study of plant communities in Switzerland, developed two concepts, one of which he called the Generic Coefficient (see below) and the other the Coefficient of Floral Community (Jaccard, 1908).

This latter was intended as a measure of relationship between two different samples, or quadrats, which he defined as 1 m². His coefficient was

$$100 \times \frac{\text{the number of species common to two quadrats}}{\text{the total number of species in the two quadrats}}.$$

The higher the value of the coefficient, the closer the similarity between the two samples. It is dependent on the number of individuals and the number of species in the communities to be compared.

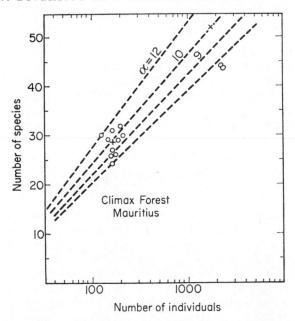

Number of individuals

FIG. 75. A series of ten samples from a climax forest in Mauritius, showing the relation between number of trees, number of species, and diversity; together with an average for the samples and the total numbers in all ten samples. The results indicate a close uniformity of the area with a diversity approximately 10.

Assuming that we are dealing with the straight-line portion of a log-series distribution, the number of species in a single quadrat containing N plant units (see p. 98) is $S_1 = \alpha \log_e N/\alpha$. If a second quadrat of the same size is taken it will have the same average number of species, but they will not be all identical with those of the previous quadrat. The two quadrats together will contain $\alpha \log_e 2N/\alpha = S + \alpha \log_e 2$. It follows that each quadrat will contain, on an average, $\alpha \log_e 2$, species not found in the other, so that the number common to the two quadrats must be $S_1 - \alpha \log_e 2$. It is interesting to note in passing that this is the number of species on half a quadrat. Thus Jaccard's coefficient of floral community is

$$100 \frac{S - \alpha \log_e 2}{S + \alpha \log_e 2} \text{ or } 100 \frac{S - 0.69\alpha}{S + 0.69\alpha}.$$

It is thus dependent on the number of species present in one quadrat (which is in turn dependent on the size of the quadrat) and on the diversity of the population. If, within the same population, the sample size is increased, the coefficient rises. If the same-sized sample is taken from a richer flora the coefficient falls. The direct measure of diversity is thus a better criterion of the relationship of floral communities, as it is uncomplicated by quadrat size.

For further discussion, see Williams (1949).

GENERIC DIVERSITY AND JACCARD'S GENERIC COEFFICIENT

Just as there are measures of specific diversity concerned with the distribution of individuals among species, so the concept can be extended to generic diversity in which the distribution of species among genera is considered. If in a community the species are spread over a large number of genera, then the generic diversity is high; if only a few genera are represented the diversity is low. The mathematical principles are the same, the species now becoming the units and the genera the groups.

Jaccard defined his "generic coefficient" as

$$100 \times \frac{\text{Total number of genera}}{\text{Total number of species}}$$

in the area sampled; or 100 times the reciprocal of the average number of species per genus, which latter he calls the "generic quotient". It has already been pointed out that the average number of units per group is dependent on sample size (pp. 7 and 256).

In the case of the log-series distribution, which many genera species distributions seem to follow, the relation between the average number of species (S) per genus (G) in a sample from a population with diversity α is given by

$$S/G = \frac{e^{G/\alpha} - 1}{G/\alpha};$$

hence Jaccard's generic coefficient is

$$100 \frac{G/\alpha}{e^{G/\alpha} - 1}.$$

In other words it is dependent on the ratio between the number of genera represented in the sample (that is to say on the size of the sample) and the richness of the population. If α remains constant, i.e. if a series of random samples are taken from the same community, the coefficient increases with sample size. If samples with the same number of genera are taken from two different communities the coefficient is simply a measure of diversity, which can be more reliably obtained from the actual numbers of genera and species than from their ratio.

For further discussion, see Williams (1949).

GENERIC DIVERSITY OF BENTHIC FAUNA IN RELATION TO DEPTH

Earlier in this chapter I have discussed the specific diversity of animals dredged at different depths in the ocean during the voyages of H.M.S. *Challenger*. Murray (1895, p. 1430) gives the number of species and of genera recorded at the same depths, and "believed to have come from the bottom" from which it is possible to estimate the generic diversity.

Table 84 gives his figures, and they are shown graphically in Fig. 72. It will be seen that there is an almost steady decrease in diversity from 779 in shallow waters to about 150 at the greatest depths. The very high values are due to the great geographical range of the stations and the great range of the animal kingdom included in the records. The generic diversity thus changes parallel to the specific diversity.

TABLE 84. *Generic diversity of benthic animals from different depths of the sea, from* Challenger's *collection*

Depth in metres	No. of stations	No. of sp.	No. of genera	Generic diversity
Under 180	70	4248	1438	779
180–900	40	1887 ·	771	490
900–1800	23	616	352	347
1800–2700	25	493	313	372
2700–3600	32	394	256	322
3600–4500	32	247	172	247
Over 4500	25	153	119	250

GENERIC DIVERSITY ON DRAGONFLIES, ODONATA, IN DIFFERENT CONTINENTS

Tillyard (1917, p. 300) gave the number of genera and species of dragonflies known at that time from different geographical regions of the world. A summary of these, together with the diversities, is shown in Table 85.

TABLE 85. *Generic diversity in dragonflies in different regions of the world*

Region	Sp.	Genera	Index of diversity
Neotropical	747	135	48
Nearctic	304	59	21
Palaearctic	256	59	23
Ethiopian	395	105	46
Oriental	595	136	55
Australian	304	110	60
Whole world	2457	429	151

It will be seen that generic diversity is low in North America, Europe and Northern Asia, distinctly higher in South America, Africa and India, which are the tropical regions, and highest of all (but possibly not significantly different from the Oriental region) in the Australian region. This latter may be partly connected with the breaking up of the land into a number of islands, or to the great range of latitude North and South. The generic diversity of the whole world is very much higher than any of the single regions into which it can be divided, indicating considerable evolutionary isolation between the regions, in spite of the strong powers of flight of many of the species.

GENERIC DIVERSITY IN AFRICAN GRASSHOPPERS

In 1957, Uvarov and Johnston published a census of the African Acridoidea in which a table was given showing the number of genera and species in different geographic regions. The seven areas included the following countries, as then recognized.

1. *Northern:* Morocco, Algeria, Tunisia, Libya, Egypt and Rio de Oro.
2. *Western:* Mauretania, Senegal, Gambia, French West Africa (inc. French Sudan), Guinea, Sierra Leone, Ivory Coast, Liberia, Gold Coast, Togo, Dahomey, Cameroons, Nigeria and Niger Colony.
3. *Eastern:* Sudan, Eritrea, Ethiopia, Kenya, Zanzibar, Somalilands, Socotra, Tanganyika, Uganda, Port. E. Africa.
4. *Central:* Fr. Equat. Africa, Fr. Congo, Gabon, Belg. Congo, Nyasaland, N. and S. Rhodesia, Angola.
5. *Southern:* Bechuanaland, S.W. Africa, Cape Province, Orange Free State, Basutoland, Transvaal, Natal, Zululand and Swaziland.
6. *Indian Ocean Islands* including Madagascar and the Seychelles (mostly very little known).
7. *Atlantic Islands:* Azores, Madeira, Canaries, Cape Verde Is. and St Helena.

Their results are condensed in Table 86 and shown diagrammatically in Fig. 76. In the table are also given the average number of species per genus for each region, and the generic diversity.

TABLE 86. *Generic diversity in Acridoidea in different regions in Africa*

Region	Genera	Sp.	Sp. per genus	Generic diversity
(1) Northern	86	284	3.30	40
(2) Western	180	446	2.47	110
(3) Eastern	241	753	3.12	130
(4) Central	241	693	2.88	130
(5) Southern	176	446	2.53	110
(6) Indian Ocean Islands	79	155	1.96	68
(7) Atlantic Islands	33	69	3.00	25
(8) Whole of Africa	501	2006	4.00	215

It will be seen that in general the regions with the largest number of species have the highest average species per genus, and also the highest diversity. The diversity for the whole area is 215, as compared with a maximum of 130 for the two tropical regions, Eastern and Southern. On the other hand, the generic diversity for the small group of Atlantic islands with sixty-nine species in thirty-three genera is only 25. The Northern Region—largely desert or semi-desert—also has a low diversity.

On Fig. 76 there has also been indicated the lines of equal numbers of species per genus, which cut across the lines of equal diversity. There is an increase in the number of species per genus associated with increase in the total number of species in the area, and with higher diversity.

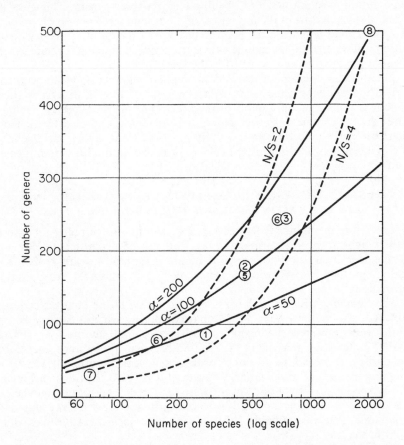

Fig. 76. Relation of numbers of genera to numbers of species (on a log scale) for Acridoidea in different parts of Africa. Also the relation between these, the diversity, and the average number of species per genera. (For number references, see Table 86.)

GENERIC DIVERSITY IN MIGRANT AND NON-MIGRANT BUTTERFLIES IN CEYLON

In the island of Ceylon there are, according to Ormiston (1917), a total of 241 species of butterflies which he classifies into 119 genera. Of these, sixty-nine species have been recorded as taking part in migratory flights, which are not uncommonly seen in this part of the world (Williams 1927, 1930). It is of interest to ask if statistical analysis supports the accepted idea that the migratory habit is more commonly associated with certain genera, rather than being randomly distributed and independent of generic classification.

The sixty-nine migrant or possible migrant species are found—using the same author's classification—to belong to forty-three genera. This gives an index of diversity, from the log series, of 48 as compared with a diversity of 91 for the total fauna. To be independent of any theory of frequency distribution, I have also calculated the values from Yule's moment formula, which gives 43 for the migrants and 98 for the total.

In either case the generic diversity among the migrants is much lower than that of the whole fauna, so that, from the point of view of generic relation, the one is not a random sample of the other. The migrant species, having a lower diversity, represent fewer genera than would be expected by a chance distribution. Therefore we have definite support for the theory that the habit of migration is found more particularly in association with certain genera than with others.

GENERIC DIVERSITY IN CAPTURES OF LEPIDOPTERA IN A
LIGHT TRAP IN ENGLAND DURING SEVEN YEARS

As previously mentioned (Chapter 3, p. 27), continuous trapping for insects with a light trap was carried out at Rothamsted Experimental Station, Harpenden, about 25 miles north of London, for the 4 years 1933–7 and again from 1946 to 1949, with the same type of trap and in the same position throughout.

The following discussion is based on the species of the family Noctuidae captured in 7 of these years (1933 omitted) during which just over 20,000 moths were captured belonging to 129 species and seventy-four genera. Table 87 shows the analysis of the relation between species and genera. For comparison there is also given the results of 4 years' captures (1955–9) at Kincraig in north central Scotland, during which 21,700 Noctuidae belonging to 121 species and sixty-one genera were captured. The greater number of individuals in a shorter time does not necessarily represent a more abundant population, and was probably due to the use of a more attractive ultra-violet lamp in the trap.

The table gives for each year, or group of years, the total number of species and genera, and the frequency distribution of genera with different numbers of species. On the assumption that the distribution closely resembles a logarithmic series, estimates have been made of the expected number of genera with only one species (n_1), and also of the generic diversity. The estimated values of n_1 are close to, but usually a little below, the observed values.

For the generic diversity it will be seen that at Harpenden single years vary from 42 to 65, but the two periods of 3 and 4 years give very similar values at 68 and 66. When these two periods are added, the diversity is slightly larger at 72. This is to be expected, as the general environmental diversity is likely to be greater over the longer period, particularly as there was a gap of 9 years between the two experiments.

TABLE 87. *Frequency distribution of genera with different numbers of species in Noctuidae (Lepidoptera) captured in S.E. England and N. Central Scotland, with diversities based on the log series*

No. of Sp.	No. of genera	Diversity	n_1 cal.	Observed no. of genera with — species						
				1	2	3	4	5	6	over six
Harpenden										
1934	75	43	43	27	27	6	5	3	2	—	—
1935	95	53	49	32	35	10	3	3	3	1	—
1936	73	49	65	34	37	6	3	1	1	1	—
1934–6	108	65	68	42	45	8	6	2	3	1	—
1946*	72	48	63	33	32	7	4	1	2	—	—
1947	81	49	52	32	33	8	2	4	2	—	—
1948	92	57	64	38	40	10	—	4	2	1	—
1949	94	54	52	34	35	10	3	1	4	1	—
1946–9	120	69	66	43	47	10	2	5	3	2	—
Total 7 years	129	74	72	46	50	11	4	5	2	3	—
Kincraig										
1955**	89	49	44	29	36	4	2	4	—	2	at 11
1956	99	54	48	32	39	3	3	5	2	1	at 9
1957	96	51	44	30	36	4	2	6	—	1	at 10
1958	101	51	39	28	36	4	1	5	1	1	at 11
1959*	95	48	37	27	33	4	1	5	3	1	at 11
1955–9	121	61	47	34	43	5	3	3	2	2	at 17 (2) and (11)

* = May to Dec. ** = Jan. to August
Total known British Noctuidae: 348 species; 148 genera; diversity 97

At Kincraig, 400 miles to the north, the generic diversities are regularly lower than in the south, ranging from 37 to 48 for single years and 47 for the whole 4 years.

For comparison with the above it should be added that the total number of species of Noctuidae in the British Isles, using the same classification, is 384, which are classified into 148 genera, and have a diversity of 97. Thus we see that the samples drawn from the total population in these two restricted localities have a distinctly lower generic diversity than the fauna of which they are a part. Also the more northerly sample, although based on a larger number of individuals, has a lower diversity than the sample from farther south.

The spread of the collections at Harpenden over 7 years gives the possibility of further analysis to see if there is any difference in generic diversity between the species which appear regularly as part of the fauna and those which are found only in some or a few of the years.

Table 88 shows the distribution of the 129 species according to the number of years, out of 7, in which they occurred and Table 89 shows an analysis of the genera–species relation in groups in descending order of frequency. On the left of Table 89 are those species which occurred in every year, next those that occurred in 6 or 7 years, in 5, 6 or 7 years, and so to those which occurred in at least 1 year, which is, of course, the total number recorded. The last column gives the species which occurred only in a single year.

TABLE 88. *Distribution of 129 species of Noctuidae according to the number of years out of 7 in which they were captured in S.E. England*

1 year only	2 years	3 years	4 years	5 years	6 years	All 7 years
17	20	10	14	12	14	42

The estimated values of n_1 (genera with one species only) are again close to the observed, but generally slightly below.

The generic diversity shows a steady increase from 30 for the very regular species which occurred in every one of the 7 years to 69 for all species which occurred in any one or more years. As an increase in generic diversity implies that the same number of species will be spread over a larger number of genera, we see that the species that occur in all or nearly all of the years, and which are presumably best adapted to the climatic vagaries of the locality, are definitely more "congeneric" than the species which only appear occasionally.

Finally it will be seen from the last column in Table 89 that the seventeen species which each occurred in only 1 year out of the 7, belonged to sixteen different genera, and give a generic diversity of about 150.

Generic diversity is therefore definitely related to types of environment, and

TABLE 89. *Genera of Noctuidae with different numbers of species captured in S.E. England, sorted according to the number of years in which the species occurred. The total was 131 species belonging to seventy-four genera*

		All 7 years	Species which occurred in at least						One year only
			6 years	5 years	4 years	3 years	2 years	1 year	
Sp. per	1 (n_1)	20	23	27	36	39	44	50	15
genus	2	3	5	7	8	8	11	11	1
	3	1	3	3	3	2	3	4	—
	4	2	1	2	1	4	3	5	—
	5	—	2	2	2	2	3	2	—
	6	—	—	—	4	1	2	3	—
Total gen.		27	34	41	51	56	66	74	16
Total sp.		42	56	68	81	93	114	131	17
Sp. per genus		1.56	1.65	1.66	1.59	1.66	1.73	1.77	1.06
Calculated n_1		17.2	21.8	26.4	34.0	36.3	41.0	45.2	15.3
Diversity (log series)		30	36	43	59	59	64	69	153

genera have climatic characteristics which can be studied mathematically. There is also evidence that the more severe or extreme is the climate the lower the generic diversity.

SEASONAL VARIATION IN GENERIC DIVERSITY

Earlier in this chapter (p. 157) I have discussed the seasonal variation in specific diversity in the Macro-Lepidoptera, as indicated by the number of individuals and species of moths caught each month in a light trap at Harpenden, in S.E. England. These showed a low diversity in spring and autumn, and a peak in the month of July.

Table 90 shows for comparison the numbers of species and of genera caught in the same trap during the years 1934–7 for the family Noctuidae only, the numbers being the totals for all repetitions of each month. On the

TABLE 90. *Comparison of seasonal changes in generic diversity in Lepidoptera captured in S.E. England and N. Central Scotland*

	Harpenden						Kincraig		
		1934–7			1946–9			1955–9	
	Sp.	Gen.	Div.	Sp.	Gen.	Div.	Sp.	Gen.	Div.
Mar.	3	2	(3)	6	3	(2.3)	9	7	(14)
Apr.	7	3	(2)	6	2	(1.1)	11	6	5
May	13	9	13	20	15	28	19	11	11
June	47	25	21	48	31	39	48	23	17
July	65	39	41	64	38	39	72	31	20
Aug.	47	28	29	62	39	45	74	40	35
Sep.	34	29	97	39	30	61	48	31	38
Oct.	16	13	31	24	20	56	25	19	37
Nov.	5	4	(10)	4	4	—	9	8	(34)

same table there are also shown similar values for the catches at Kincraig, Inverness-shire (about 400 miles north of Harpenden) in the years 1955–9. It should be noted that all diversities calculated from less than ten species are subject to a very high error.

An examination of the table and of Fig. 77 shows the following points. In both localities there is a low diversity in early spring rising steadily to midsummer. At this point, however, the parallel with specific diversity breaks down, as the Kincraig results show the diversity rising slowly to a peak in September and October, while at Harpenden, after a slight fall in August, the diversity suddenly rises to an extreme of 90, when out of twenty-nine genera represented only five had two species each and no genus had more than this. After this month there was, at Harpenden, a rapid fall to low values in late autumn.

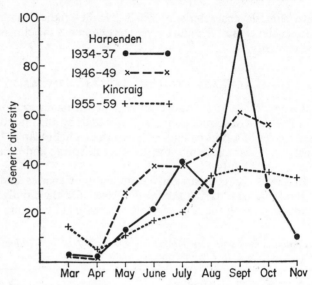

FIG. 77. Seasonal changes in generic diversity of Lepidoptera caught in light traps at Harpenden (S.E. England) and at Kincraig (North Central Scotland).

THE RELATION OF GENERIC DIVERSITY TO HABIT IN THE BRITISH BIRD FAUNA

In the list of Birds of Great Britain and Ireland published in 1952 by the British Ornithological Union, each of the 426 species is given briefly a status, such as "straggler", "resident", etc. In some cases a single species, or some of its recorded sub-species, come into more than one category.

Table 91 and Fig. 78 show first the frequency distribution in genera of all the recorded species, and then of those in six groups: stragglers, irregular visitors, passage migrants, non-breeding migrants, breeding migrants and residents. As one species may occur in more than one category it is not possible to add the numbers in groups together to get a larger group.

The table shows in addition the number of genera and species in each category, and the average number of species per genus. Also, calculated from these on the assumption of the log-series distribution, are estimates of x, n_1 and the generic diversity for the birds of each habit. In Table 92 the categories are listed again in descending order of diversity.

Taking the total fauna of all orders together it will be seen that the generic diversity of all the British species is 138, while that of the resident species alone is a little lower at 120. The highest diversity is found in the "irregular visitors", where it is 175, probably reflecting the very great area from which such birds may be drawn. On the other hand, the rarer "stragglers", with only a few recorded individuals, have a much lower diversity, 95, indicating a

TABLE 91. *Distribution of genera with different numbers of species in various habit groups of British birds, with special relation to migration*

	Number of genera with — species								and at	Total genera	Total species	Species per genus	"x"	"n_1"	Generic Diversity
	1	2	3	4	5	6	7	8							
All British birds															
Non-Passerine	79	25	6	9	2	1	2	1	9, 11, 12 (2), 13	129	265	2.05	.73	71.6	98
Passerine	36	9	3	2	3	2	3	—	9, 10 (2), 13	62	161	2.60	.814	30.0	37
Total	115	34	9	11	5	3	5	1	} 8	191	426	2.23	.765	100.0	131
Stragglers															
Non-Passerine	45	5	2	3	1	1	—	—		57	84	1.47	.52	40.3	77
Passerine	24	3	6	2	2	1	—	1		39	80	2.05	.73	21.6	30
Total	69	8	8	5	3	2	—	1		96	164	1.71	.632	60.3	95
Irregular visitors															
Non-Passerine	42	5	3	—	—	—	—	—		50	61	1.22	.317	41.7	131
Passerine	16	5	1	—	—	—	—	—		22	29	1.32	.405	17.3	43
Total	58	10	4	—	—	—	—	—		72	90	1.25	.34	59.4	175
Passage migrants															
Non-Passerine	20	9	—	2	—	2	—	—		33	58	1.76	.65	20.3	31
Passerine	11	1	—	—	—	—	—	—		12	13	1.08	.15	11.1	(74)*
Total	31	10	—	2	—	2	—	—		45	71	1.58	.575	30.2	53
Not breeding migrants															
Non-Passerine	26	6	6	1	2	—	—	—		41	70	1.71	.634	25.6	40
Passerine	10	2	—	—	1	—	—	—		13	19	1.46	.50	9.5	19
Total	36	8	6	1	3	—	—	—		54	89	1.65	.61	34.7	57
Breeding migrants															
Non-Passerine	27	3	1	—	1	—	—	—		32	41	1.28	.37	25.8	69
Passerine	15	1	2	1	—	—	—	—		19	27	1.42	.48	14.0	29
Total	42	4	3	1	1	—	—	—		51	68	1.33	.415	39.8	96
Residents															
Non-Passerine	41	7	4	1	1	—	—	—		54	76	1.41	.473	40.0	85
Passerine	24	3	—	2	1	1	—	—		31	49	1.58	.577	20.7	36
Total	65	10	4	3	2	1	—	—		85	125	1.47	.51	61.3	120

*Very high error (see text).

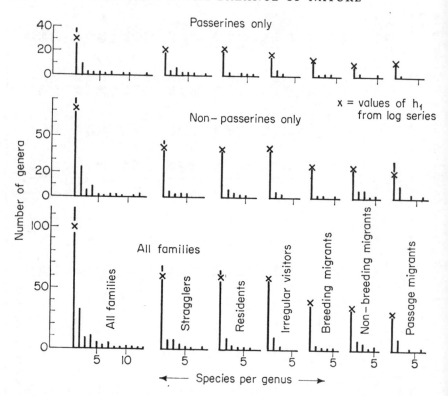

FIG. 78. Frequency distribution of genera of British birds with different numbers of species: the birds grouped according to their habits as resident or migrant, and also separated as passerine or non-passerine.

TABLE 92. *Different habit-groups of British birds in descending order of generic diversity*

| Group | No. of sp. in group | Generic diversity | | |
		Passerine	Non-passerine	All orders
Irregular visitors	90	43	131	175
All birds	426	37	98	138
Residents	125	36	85	120
Stragglers	164	30	77	95
Breeding migrants	68	29	69	96
Non-breeding migrants	89	19	40	57
Passage migrants	71	74*	31	53

* This value is subject to a very high error (see text).

higher proportion of congeneric species. The more regular visitors, both breeding and non-breeding, come next in the series, with the latter slightly less diversified, while the "passage migrants" have the lowest diversity, 53, and hence the highest proportion of congeneric species.

When the species are further separated into the 161 belonging to the passerines and the 265 in the remaining eighteen orders, the sequence of diversity in relation to habit is identical in each group suggesting a real biological significance. The abberrant value of 74 for the passerine passage-migrants is calculated from thirteen species in twelve genera and is subject to a very large error. Had there, in fact, been one more genus the diversity would have been infinite, and with one less genus it would only have been 35.

The estimated values of the number of genera with one species on the assumption of the log series have been added to Fig. 78 and the close fit to the observed values will again be seen.

GENERIC DIVERSITY OF LEAFHOPPERS IN FINLAND

Previously (p. 161) I have discussed some data provided by Kontkanen (1950) on the distribution of species of leafhoppers (Homoptera) in Finland.

In Table 73 (p. 162) and Fig. 79 are shown the number of genera and species which he found in thirty different biotypes in North Karelia. On the figure are also shown the expected relations for different generic diversities on the basis of a log-series distribution. In the biotypes with fewer than about twelve species the calculation of diversity is subject to a very high error, and

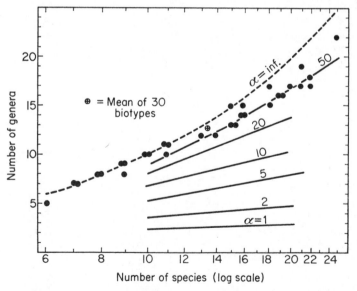

FIG. 79. Generic diversity of leafhoppers in thirty different biotypes in North Karelia, Finland (Kontkanen, 1950).

the number of cases with low numbers of species and an "infinite" diversity are those in which each species is in a different genus.

It will be seen that, apart from these, there is a general clustering of the biotype diversities near or just above the line indicating a diversity of 50. The average of all biotypes, from the arithmetic mean of the genera and the mean log of the species, is also shown in the diagram. The generic diversity of the whole area covered by the thirty biotypes (eighty-seven species in fifty-two genera) was 53.4.

Chapter 8

THE FREQUENCY DISTRIBUTION OF PARASITES ON OR IN THEIR HOSTS

THE pattern of the frequency distribution of parasites on the host animals or plants seems to have been neglected by biologists until quite recently.

Before discussing some of the evidence for a mathematical order in such distributions it is important to point out once more that the method of obtaining information by sampling methods with parasites and their hosts is nearly always basically different from that used in the study of individuals and species. In the latter it is usual to take a random sample of individuals (units) and then to see to what species (group) each one belongs. In the case of the host–parasite relationship it is the normal procedure to take a random sample of hosts (groups) and to see how many parasites (units) are on each.

This has a very important mathematical effect when one comes to compare the original population with samples taken by the two methods. In the first case, increasing the size of the sample, by getting more units, will normally add to the number of units in groups already represented, as well as adding new groups. For example, in the case of individual animals and species, the addition of more individuals to the sample causes some or many of the species to move up to higher classes of abundance (number of individuals per species). When sampling by groups no increase in sample size can alter the number of units in any of the groups already examined. In the first case, the average number of units per group increases with increasing sample size; in the second, additional sampling, if random, should make no difference to this average.

It follows that, in the case of a logarithmic series distribution, increased sampling by units increases the average number of units per group and hence the value of "x", but leaves the diversity "α" the same as that of the population. Increased sampling by groups, however, leaves the average number of units per group, and hence the value of "x", constant and the same as that of the population, but increases the value of the diversity. This cannot, however, become greater than that of the population sampled.

From a biological point of view the distribution of parasites on hosts is a very complex problem with many different factors influencing the result.

For example, there are cases where the average number of parasites per host is always low, particularly when the size of the parasite is not very small when compared with that of the host. In some of the cases discussed below the average infection is only just above one per host: in the case of the blood-sucking Hippoboscidae on birds (p. 225) the highest infection is ten on a single host. The small number of classes into which such a low infection can be grouped makes mathematical study difficult. At the other end of the scale we have the malaria parasites, and in the example quoted on p. 243 the number reaches 134,000 in a tiny sample of blood from one child, with a total infection of perhaps over a 1000 million.

Parasites may be external, as in the Hippoboscid flies, fleas and mites, or internal as in the malaria parasites, Filaria and Cestodes. A complex case is that of the Copepods parasitic on mussels (p. 230) which although within the shell and the mantle cavity of the host are not within its body cavity. The hazards of life are very different in these contrasting situations.

The host animal may be selected (within certain limits) by the parasite itself, as for example with the fleas and Hippoboscids. On the other hand, parasite distribution may be determined at a different stage in the life cycle, as when the female Ichneumon seeks out particular caterpillars on which to lay eggs. In this case the frequency distribution of the larvae will depend on the egg-laying habits of the parent, whether they lay singly or in numbers at a time in a single host. The selection of host in the case of the malaria or filaria parasites may be made by an entirely different species—the insect transmitter. Or the host distribution may be almost by chance, depending even on the habits of the host, as when the eggs of a parasite are widely distributed on the possibility that some may be taken up by the host either by contact or through the mouth.

Some parasites remain associated with a single host, not merely through their lifetime, but also by breeding on it through successive generations, so that the population on the host multiplies without reinfection: this is the case, for example, with the lice and the mites. Others, such as fleas and Hippoboscidae in the course of their life may feed on several hosts, but without breeding on any host. In both these groups it is only the adult that is blood-sucking. The larval fleas feed on blood-contaminated refuse, often within the nest of the host bird or mammal. The larval Hippoboscid develops entirely within the body of its parent, and when laid is already full-fed and ready to pupate (pupiparous). In the lice and the mites all stages feed on the host.

By a slight extension of definition we can consider plant-eating insects as parasites of the plants on which they feed, and here it is possible to vary the size of the host group from a single leaf to a whole tree. In the case of an insect which is only thinly distributed, if a single leaf is taken as the group the number of zero values may be excessive and may make analysis difficult. By making the group a specified number of leaves or a whole plant, the number of zeros can be greatly reduced or even eliminated, as mentioned in the case of scale insects on Citrus leaves on p. 249. The mathematical pattern behind the distribution is not affected by the size of the group.

The distribution of insects on the leaves of plants is affected by the oviposition habits of the parent and by the mobility of the parasites themselves. If the eggs are originally laid in batches some form of contagious distribution will be found, which will, however, be obscured by the mobility of the parasites after hatching. Rarely there may be a reverse effect due to gregariousness in some caterpillars.

DISTRIBUTION OF LICE ON HUMAN HEADS

In a series of papers between 1936 and 1941, P. A. Buxton discussed some problems of the distribution of head-lice (*Pediculus humanus capitis;* Anoplura.) on the heads of human beings, dealing in 1940 particularly with information obtained from Southern India. Later he handed the original data over to me, for further study (see Williams, 1944).

It is important to note, before discussing these, that the method of counting the lice led to an unusually high accuracy even with the smaller numbers. The material was obtained from a jail in Cannamore where the prisoners' heads were shaved for sanitary purposes on arrival. By arrangement with the authorities each head of hair was kept separate in a closed bag, and later treated with a solution of sodium sulphide, which dissolved the keratin of the hair, but not the chitin of the insects. The lice were then washed free from the residue by the use of very fine sieves.

The data for male prisoners are shown in Table 93. The total number of heads examined was 1083, of which 622 were free from lice. On the remaining 461 there were 7442 lice (including all stages except eggs) giving an average of

TABLE 93. *Frequency distribution of head-lice* (Pediculus) *on human heads in India. 7442 lice distributed on 461 heads out of a total of 1083 examined. The remaining 622 were free from lice*

Lice per head	Obs.	Log ser.	Neg. bin.	Lice per head	Obs.	Log ser.	Lice per head	Obs.	Log ser.
0	622	—	622*	17	7	5.0	34	—	1.9
1	106	107.2	86.3	18	4	4.7	35	2	1.9
2	50	52.8	48.3	19	7	4.3	36	2	1.8
3	29	34.7	33.8	20	7	4.1	37	4	1.7
4	33	25.6	26.0	21	3	3.8	38	3	1.7
5	20	22.2	21.2	22	4	3.6	39	—	1.6
6	14	16.6	17.8	23	4	3.4	40	4	1.5
7	12	14.0	15.3	24	4	3.2	41	1	1.5
8	18	12.1	13.4	25	3	3.0	42	—	1.4
9	11	10.6	11.9	26	4	2.9	43	—	1.4
10	11	9.4	10.7	27	6	2.7	44	2	1.3
11	3	8.4	9.7	28	2	2.6	45	3	1.3
12	10	7.6	8.8	29	4	2.5	46	1	1.2
13	8	6.9	8.1	30	1	2.3	47	5	1.2
14	6	6.3		31	—	2.2	48	1	1.1
15	3	5.8		32	1	2.1	49	—	1.1
16	6	5.4		33	5	2.0	50	1	1.1

Also single heads with the following numbers of lice:
51, 53, 54, 57, 58, 59, 60, 64, 66, 71, 74, 79, 80, 83, 88, 110, 121, 128, 129, 145, 149, 188, 239, 270, 303, and 385.

*by assumption

16.2 lice per infested head. This arithmetic mean was not, however, the median, nor was it near the peak of frequency. This peak was at 106 heads with only a single louse each; fifty heads had two lice, and so on in diminishing numbers as shown in Fig. 80. The most heavily infested head had 385 lice, which is 5 per cent of the total. The eleven heads each with over 100 lice accounted for nearly 29 per cent of the total lice; the 106 least-infested heads accounted for only 1.4 per cent of the total lice. In fact, the distribution is definitely of the "hollow curve" type with the majority of the groups with a few units, and the majority of the units in a few groups.

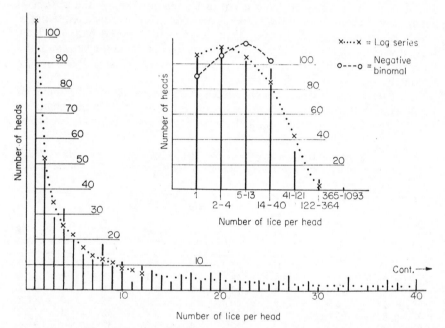

FIG. 80. Frequency distribution of lice on heads of man, showing a close relation to the logarithmic series (dotted line). Inset are the same data in × 3 log classes of abundance. (From *J. Anim. Ecol.* (1954), **42**, 9.)

If we omit the heads with no lice, it is possible to calculate a logarithmic series for the 7442 lice on 461 heads which gives values of $x = .9856$ and $\alpha = 108.7$. The terms of this series from one to fifty lice per head are shown in Table 93 for comparison with the observed numbers.

The data, both for the original and the log series, are shown grouped in × 3 log classes in Table 94, and also in the diagram inset in Fig. 80. It will be seen how extremely closely the calculation fits to the original data, especially for the first three classes. The calculated number of heads with one louse (n_1) is 107, with an observed value of 106. For heads with two to four lice the calculated number is 113, the observed 112; and for five to thirteen lice per head, calculated 107.9, observed 107. A closer fit is almost impossible. However, on calculating the higher values Class IV (14–40) gives somewhat too few, and Class V (41–121) too many, leaving rather fewer very heavily infested heads in the calculated than in the observed figures.

Although the fit of the log series is so close it was desirable to test other alternatives, so a negative binomial (which includes the zero term 622 as a parameter) was calculated with $k = 0.1415$ and $p = 48.59$, with the results as shown in Tables 93 and 94, and in Fig. 80. The calculated number of heads with one louse is a poor fit at eighty-six instead of 106. The fit of Classes II and III is closer, but IV is again poor.

TABLE 94. *Frequency distribution of heads with lice, classified in × 3 log classes*

Class		Obs.	Log ser.	Negative binomial
I	(1)	106	107.2	86.3
II	(2–4)	112	113.1	108.1
III	(5–13)	107	107.9	116.7
IV	(14–40)	96	84.1	102.2
V	(41–121)	31	43.3	
VI	(122–364)	8	} 5.4	
VII	(365–1093)	1		

As the distribution of the × 3 classes in Fig. 81 resembles a truncate log normal, this type of frequency was investigated, by means of a log-probability graph, from the accumulated percentages of the total species up to the end of each class. In this case a term for the heads without lice is desirable, and it is necessary to decide if the 622 heads without lice in the original sample are representative of the population from which the infested heads came, or whether some at least of the free heads might come from social classes in which the majority of heads would be uninfested (see p. 7). The shape of the truncate curve in Fig. 81 suggests that the peak, and part below the peak, is shown among the infested heads. If the complete curve were symmetrical the area below Class I should be approximately equal to the number of heads in Class IV and above. This is actually 138, considerably less than the observed 622.

As there is no firm basis for deciding this point, calculations were made for three values of heads without lice, 622, 139 and omitting it completely. The accumulated percentage totals of heads in each case is shown in Table 95 and the resulting graphs in Fig. 81.

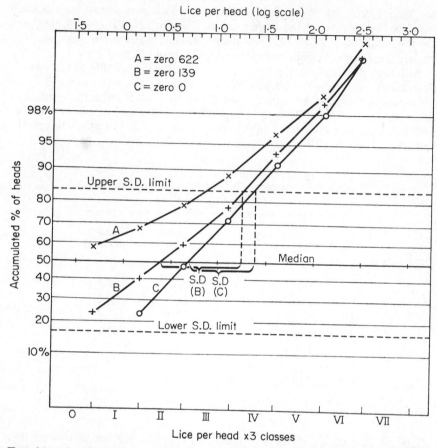

FIG. 81. Distribution of head-lice on man. Relation between the accumulated percentage of heads (on a probability scale) and the number of lice per head on a logarithmic scale. A, with zero term = 22 as observed; B, with zero term estimated at 139; and C, neglecting the zero term. The closest relation to a straight line, indicating a log-normal distribution, is obtained by omitting the zero term.

It will be seen that, unexpectedly, the nearest approach to a straight line is when the zero term is omitted. The line in this case indicates a log-normal distribution with a median of approximately 0.7 and a S.D. of 0.7 on the $\log_{10}$ scale. The omission of all the zero heads is, however, doubtfully justified, as it is impossible to believe that there were no free heads in the population from which the infested heads were drawn. On the assumption of a symmetrical distribution with the zero term 139 the curve is moderately close to a straight line with a median at 0.4 and a S.D. of 0.65, but not as good a fit as the first. The line of the values including all the 622 free heads is definitely not straight.

We can therefore suggest from the analysis that the frequency distribution of infested heads with different numbers of lice is very close indeed to a logarithmic series, particularly in the lower infestations. It is a poor fit, particularly at low infestations, to a negative binomial, and also not a good fit to a log-normal except by making some assumptions which are not entirely justifiable biologically.

TABLE 95. *Accumulated percentage of total heads with different level of infection by lice; with different zero terms considered*

Class 0, no. lice =	622	139	0
	%	%	%
Class 0	58	23.2	0
I	68	40.8	23.0
II	78.5	59.5	47.4
III	89	77.3	70.7
IV	96.2	93.3	91.5
V	99.0	98.5	98.0
VI	99.9	99.8	99.8

FLEAS ON RATS IN BURMA

Some years ago Dr J. L. Harrison sent me data, collected in Rangoon, Burma, in 1945, on the occurrence of fleas on several species of rats (*Bandicota* and *Rattus*) and on a shrew mouse (*Suncus*). Unfortunately the two species of fleas present, *Xenopsilia astria* Roths. and *X. cheopis* Roths., were not counted separately.

The frequency distribution on the different hosts is given in Table 96 and Fig. 82: they are of the familiar hollow-curve type. The average number of fleas per host is low, ranging from 3.11 in Bandicota to 1.41 for the shrew mouse, the difference being possibly partly connected with the size of the animals.

Table 97 shows the first five terms of the log series calculated to fit the total infestation (omitting those free from fleas) on each of the five hosts. It will be seen that the calculated n_1 is slightly too high in *Bandicota*, a little too low in the three species of *Rattus*, and almost exact in *Suncus*. In the animals with two to four fleas the results are almost the reverse; too low in Bandicota, too high in *R. norvegicus* and *R. rattus*, and nearly identical in *R. concolor* and *S. coeruleus*.

TABLE 96. *Frequency distribution of small mammals in Burma with different numbers of fleas (two species not separated)*

Host	Bandicota bengalensis	Rattus norvegicus	Rattus rattus	Rattus concolor	Suncus coeruleus
Total no. examined	521	121	111	387	87
Without fleas	242	74	76	270	70
As % of total	46.4	61.1	68.5	69.8	80.5
Fleas per host					
1	116 (116)	27 (27)	26 (26)	68 (68)	13 (13)
2	53	10	3	21	2
3	38	3	2	14	1
4	28 (119)	— (13)	1 (6)	5 (60)	1 (4)
5	7	2	1	1	—
6	9	2	—	4	—
7	5	1	1	2	—
8	3	—	1	1	—
9	4	1	—	—	—
10	3	—	—	—	—
11	—	—	—	1	—
13	3 (34)	— (6)	— (3)	— (9)	— (0)
14	3	1	—	—	—
15	4	—	—	—	—
18	1	—	—	—	—
20	1	—	—	—	—
27	1	—	—	—	—
Total rats with fleas	279	47	35	117	17
Total fleas	868	108	62	234	24
Aver. fleas per rat with fleas	3.11	2.30	1.77	2.00	1.41
Aver. fleas all rats	1.67	0.89	0.56	0.60	0.28

For *B. bengalensis* the log series was calculated to the 13th term as shown diagrammatically in Fig. 83, both for each term and the × 3 log classes. The calculated values are very close for Class I, too low for II, too high in III and close, but in small numbers, for animals with over thirteen fleas.

For the same host a calculation was also made to test the fit to a negative

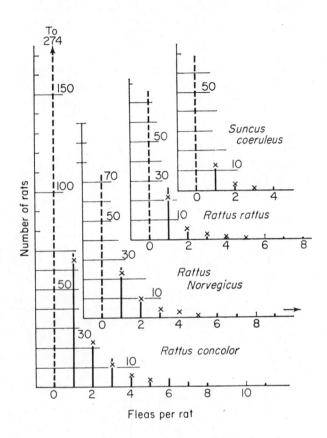

FIG. 82. Distribution of fleas on four species of rats in Burma with values (x) calculated from the logarithmic series for the earlier terms.

binomial, as seen in Table 97. Here the zero term is included and is correct by assumption. The value for n_1 is too small and for n_2 too large, both with poorer fits than the log series; n_3 is almost correct and n_4 too small but both closer than the log series.

It is difficult to draw any definite conclusions from these and the situation may be made more complex by the presence of the two species of flea.

Finally, also for *B. bengalensis*, Fig. 84 shows the accumulated percentages plotted on probability paper, showing that these fit very closely to a straight line. This indicates that the frequency is very close to a log-normal distribution with a median at $\bar{1}.9$ (0.8 fleas per rat) and a S.D. of about 0.56 on a $\log_{10}$ scale. The observed number of rats without fleas is slightly higher than that indicated by the extrapolation of the straight-line relation.

TABLE 97. *Distribution of small mammals in Burma with different numbers of fleas, the infection being measured in × 3 log classes*

Fleas per rat	*Bandicota bengalensis* Obs.	Calc. log series	neg. binomial
0	242	—	242*
1	116 (116)	122.3 (122.3)	100.2 (100.2)
2	53	52.5	58.4
3	38	30.1	37.3
4	28 (119)	19.4 (102)	24.9 (120.6)
5–13	(34)	(47.8)	
over 13	(10)	(7.1)	

Fleas per rat	*Rattus norvegicus* obs.	calc.	*Rattus rattus* obs.	calc.	*Rattus concolor* obs.	calc.	*Suncus coeruleus* obs.	calc.
1	27	23.3	26	21.6	68	64.5	13	12.6
2	10	9.4	3	6.0	21	23.2	2	3.0
3	3	4.9	2	2.2	14	11.1	1	0.95
4	0	2.8	1	1.5	5	6.0	1	0.34
2–4	13	17.1	6	9.7	40	40.3	4	4.3
5	2	1.8	1	0.8	1	3.5	0	0.13
1–5	42	43.2	33	32.1	109	108.3	17	17.0

* By assumption.

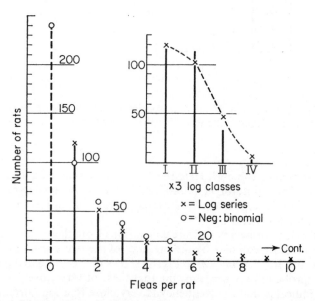

FIG. 83. Frequency distribution of *Bandicota bengalensis*, in Burma, with different numbers of two species of fleas. Number of fleas per rat is given on arithmetic and log × 3 scales.

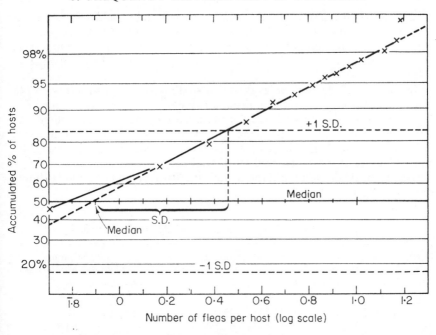

FIG. 84. Distribution of two species of fleas on *Bandicota bengalensis* in Burma, showing the accumulated percentage of rats, on a probability scale, plotted against the number of fleas per rat on a logarithmic scale. The straight-line relation indicates a close fit to a log-normal distribution.

FLEAS ON SMALL RODENTS AT OXFORD, ENGLAND

In 1950, Evans and Freeman published some results of trapping small mammals in the neighbourhood of Oxford, from which the fleas were taken and recorded separately for each host. The catches were made on 4 days in each of 10 months, from May 1938 to April 1939, omitting July and August.

The commonest flea was *Ctenophthalmus agyrtes nobilis* (Roths.), and it was found chiefly on two rodents, the Long-tailed Wood Mouse, *Apodemus s. sylvaticus* (Linn.) of the family Muscidae, and the Bank Vole, *Clethrionomys glareolus britannicus* (Miller) of the family Cricetidae.

Out of 730 wood mice examined 465 were without fleas, while on the remaining 265 mice there were 605 fleas. Out of a total of 325 bank voles, 177 were without fleas, while the remaining 148 had 417 fleas. The frequency distributions of the number of fleas per host is shown in Table 98 and in Fig. 85.

The distribution is not Poisson, nor does it give a fit, when tested on probability paper, to a log-normal distribution. On the other hand, there is a very close relation to the logarithmic series as shown in the same table. In the case of the wood mice a series of this type with 605 units (fleas) in 265 groups (hosts with fleas) gives $x = 0.771$ and calculated values for the hosts with one, two, three and four fleas as 139.1, 53.6, 27.5 and 15.9 as compared with the observed values of 138, 47, 34 and 19. The numbers for the first three $\times$ 3 log classes are 139.1, 97 and 28.5 as compared with the observed 138, 100 and 26.

In the case of the bank voles the calculated log series for 417 fleas on 148 voles with fleas gives $x = 0.836$. From this the number of hosts with one flea should be 68.4 as compared with an observed 69; and with two, three and four fleas 28.6, 15.9 and 10 as compared with 26, 15 and 11, only once differing by more than a single host. The three $\times$ 3 log classes give 68.4, 54.5 and 22.7 as against the observed 69, 52 and 26.

There is little doubt that in this set of observations the fit to a logarithmic series is extremely close.

TABLE 98. *Distribution of fleas on small mammals at Oxford*

Host	Apodemus s. sylvaticus		Clethrionomys glareolus britannicus	
No. of hosts	730		325	
No. without fleas	465		177	
As % of total	63.7		54.5	
Fleas per host	Observed	Log series	Observed	Log series
1	138 (138)	139.1 (139.1)	69 (69)	68.4 (68.4)
2	47	53.6	26	28.6
3	34	27.5	15	15.9
4	19 (100)	15.9 (97.0)	11 (52)	10.0 (54.5)
5	10	9.8	7	6.7
6	8	6.3	7	4.7
7	1	4.2	5	3.3
8	1	2.8	1	2.4
9	1	1.9	2	1.8
10	2	1.3	—	1.4
11	3	0.9	3	1.0
12	—	0.7	—	0.8
13	— (26)	0.5 (28.5)	1 (26)	0.6 (22.7)
17	—		1	
20	1		—	
Hosts with fleas	265		148	
Total fleas	605		411	
Average fleas per host with fleas	2.28		2.82	
Average fleas per all hosts	0.89		1.26	

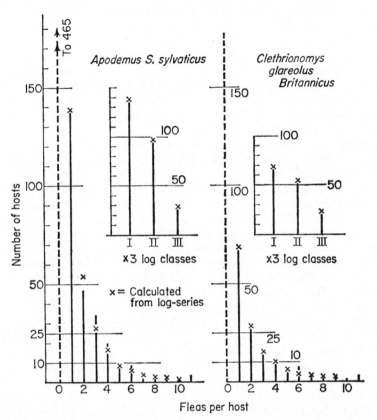

Fleas per host

FIG. 85. Frequency distribution of fleas on small mammals near Oxford, England, showing values (x) calculated from the logarithmic series which are very close to the observed numbers.

FLEAS ON RABBITS IN SCOTLAND

In 1956, Allan published a short account of the distribution of the flea, *Spilopsyllus cuniculi* (Dale), on wild rabbits on an estate in Aberdeenshire, Scotland, over a period of about 15 months. The information was obtained by trapping ten male and ten female animals every 2 weeks, and counting separately the number of male and female fleas on each. Trapping started on 24 December 1953 and was continued till March 1955, but at this point, owing to the rapid spread of myxomatosis in the district, the survey had to be abandoned, as it was no longer possible to get the required number of hosts.

Some details of the frequency distribution were discussed in this article, and Dr Allan kindly gave me a copy of the original records for each individual rabbit. Unfortunately, however, the figures for periods 25 and 26 had been mislaid and so only an approximate value for the mean log number of fleas per host is available for these two periods, from a figure in Allan's publication.

These records are of particular interest in comparison with the cases already discussed, owing to the much higher level of infestation, possibly due to the larger size of the host, and also for the spread of the observations over a complete year with very definite seasonal changes. The figures show average infestations, for the twenty rabbits in each 2-week period, ranging from 3.2 to 163 fleas per host, according to the season, and from 0 up to 430 fleas on single rabbits. In the previously considered examples the highest infestation was just over three fleas per host.

Table 99 shows Allan's data summarized for each period, giving first the number of rabbits without fleas, then the number with fleas at different levels of infestation in $\times$ 3 log classes; also the total number of fleas and the arithmetic mean per rabbit (including all rabbits), and the mean log $n + 1$ with the standard deviation of the 20 values. Figure 86 shows diagrammatically the seasonal changes in the mean log together with its error (which in this case equals the S.D. $\div$ 4.36), the Standard Deviation, and the extreme limits of infestation on a single host. It will be seen that the level was low (mean log about 0.7) at the start of the investigation at the end of December, but rose rapidly to a level of about 100 fleas per rabbit in March and April. At the beginning of May the peak infestation (on a log scale) was reached with a mean log $n + 1$ of 2.18, and with an extremely low S.D. and total range. After this there was a very rapid fall to mean logs of about 1.0 (about one-tenth of the previous level) and with a great increase in variability. In period 14, for example, the range of fleas per rabbit in the twenty animals was from 1 to 265, as compared with 114 to 208 at the peak in period 10. The rapid decrease in variability from period 6 to 10, with the level high in all, is remarkable and requires further study to see if there is any biological explanation. The absence of records for the following season for comparison was, in more ways than one, a great tragedy.

The rise in the 2nd year is much earlier, and a level of about 100 fleas per host at the beginning of January 1955 can be compared with a level of about ten a year previously. It is possible that this may be due to the rapid extermination of the rabbits by disease without any immediate reduction in the number of fleas.

The frequency distribution of all fleas on all rabbits in the first 11 months is shown diagrammatically in Fig. 87. It will be seen that, apart from the thirty-six rabbits without fleas, those with fleas give a distribution which is definitely not of the log-series form. Although the majority of rabbits have few fleas, there is not the steady fall from a peak at rabbits with one flea which is characteristic of this distribution.

Table 99 shows the number of rabbits with different numbers of fleas in each $\times$ 3 class, in each of the periods, and also totals for the first twenty-four periods. These latter are shown graphically in Fig. 88A, and it will be seen that, excluding the zero term, it closely resembles a log-normal type of distribution. This resemblance is brought out even more clearly in Fig. 88B, where the accumulated totals, as percentages of the total rabbits with fleas, are plotted

TABLE 99. *Distribution of fleas on twenty rabbits at 2-week intervals near Aberdeen in Scotland*

2-week period	First date	0	I	II	III	IV	V	VI	VII	Total fleas on 20 hosts	Arith. mean	Mean log and S.D.
		Number of fleas per rabbit × 3 class									Fleas per host	
1	1953 28 Dec.	4	2	7	2	4	1	—	—	208	10.4	0.70 ± 0.57
2	1954 8 Jan.	2	2	6	8	2	—	—	—	125	6.3	0.72 ± 0.38
3	22 Jan.	—	—	—	5	11	4	—	—	541	27.1	1.39 ± 0.26
4	5 Feb.	—	1	0	2	7	9	1	—	927	46.3	1.53 ± 0.44
5	19 Feb.	—	—	1	9	9	1	—	—	351	17.5	1.20 ± 0.26
6	5 Mar.	—	—	—	3	3	1	12	1	3264	163.2	2.01 ± 0.52
7	19 Mar.	—	—	—	—	—	13	6	1	2573	128.7	2.03 ± 0.26
8	2 Apr.	—	—	—	—	—	13	7	—	2145	107.3	2.00 ± 0.19
9	16 Apr.	—	—	—	—	—	15	5	—	1832	91.6	1.94 ± 0.15
10	30 Apr.	—	—	—	—	—	2	18	—	3054	152.7	2.18 ± 0.07
11	14 May	—	—	3	3	6	7	1	—	725	36.3	1.39 ± 0.43
12	28 May	2	2	2	6	5	3	—	—	390	19.5	1.00 ± 0.57
13	11 June	—	—	6	8	5	1	—	—	247	12.3	1.01 ± 0.32
14	25 June	—	1	4	8	4	1	2	—	639	31.9	1.09 ± 0.54
15	9 July	3	3	3	7	2 '	1	1	—	350	17.5	0.77 ± 0.59
16	23 July	8	4	4	3	1	—	—	—	64	3.2	0.40 ± 0.43
17	6 Aug.	9	2	3	4	2	—	—	—	67	3.3	0.41 ± 0.45
18	20 Aug.	2	0	3	8	6	1	—	—	267	13.3	0.96 ± 0.45
19	3 Sep.	3	1	4	4	6	2	—	—	249	12.5	0.87 ± 0.53
20	17 Sep.	—	3	4	2	10	1	—	—	267	13.3	0.96 ± 0.46
21	1 Oct.	2	0	4	10	4	—	—	—	220	11.0	0.92 ± 0.41
22	15 Oct.	1	0	3	9	7	—	—	—	223	11.1	0.98 ± 0.35
23	29 Oct.	—	—	—	10	9	1	—	—	345	17.3	1.17 ± 0.29
24	12 Nov.	—	—	1	1	9	7	2	—	1006	50.3	1.56 ± 0.39
25	26 Nov.	—	—									*1.81
26	10 Dec.											*1.36
Year total		36	21	58	112	112	84	55	2	21,779	41.9	

* Estimated from Allan, 1956, Fig. 1.

2-week period	Starting date	No. fleas	I	II	III	IV	V	VI	VII	Total fleas on 20 hosts	Arith. mean	Mean log $n+1$ and S.D.
			Fleas per rabbit × 3 classes								Fleas per host	
1	1954 24 Dec.	—	—	—	—	1	11	8	—	2011	100.5	1.97 ± 0.20
2	1955 7 Jan.	—	—	—	2	5	9	4	—	1631	81.5	1.75 ± 0.42
3	21 Jan.	—	—	—	—	3	5	12	—	2775	138.7	2.05 ± 0.35
5	4 Feb.	—	—	—	—	—	8	12	—	3085	154.3	2.13 ± 0.24
4	16 Feb.	—	—	—	1	3	6	9	1	2917	145.3	2.03 ± 0.41
6	4 Mar.	—	—	—	—	3	7	8	2	3145	157.3	2.05 ± 0.41

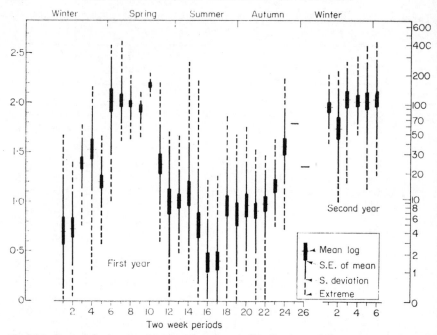

FIG. 86. Seasonal changes in abundance and distribution of fleas on samples of twenty rabbits taken each 2 weeks in Scotland for just over 1 year. The different forms of the vertical line indicate the actual mean, the standard error of the mean, the standard deviation, and the upper and lower limits of numbers on a single rabbit in the sample (data from Allan). The numbers of fleas per rabbit are on a logarithmic scale.

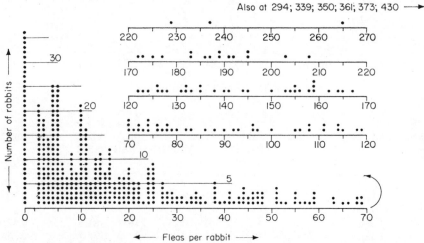

FIG. 87. Frequency distributions of 480 rabbits with different numbers of fleas taken during 1953 in Scotland, with number of fleas per rabbit on an arithmetic scale. Periods 25 and 26 not included.

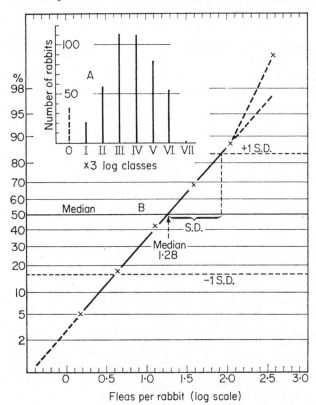

Fleas per rabbit (log scale)

FIG. 88. Distribution of fleas on rabbits. The data from Fig. 87 shown with × 3 classes of abundance. Also the same data plotted as accumulated percentage of rabbits, on a probability scale, against number of fleas per rabbit on a logarithmic scale. Both diagrams suggest a very close relation to a log-normal distribution with a median at about twenty fleas per rabbit (1.28 on the log scale).

against a probability scale, and give a very close fit to a straight line indicating a log-normal distribution with a median at 1.28 on the $\log_{10}$ scale, and a S.D. of 0.67.

There is a departure in the highest class indicating too few rabbits with over 364 fleas in Class VII, but here the scale is so open that an error in position of a single rabbit has quite a considerable effect. As already pointed out, any error in the collecting of parasites from hosts is always a reduction. In the present case the fleas were removed with a pair of forceps from the host in the field as soon as it was killed. Some part at least of the excess of zero counts and the shortage of the highest infestations might well be due to overlooking a single flea, or to the escape of a few fleas out of two or three hundred on a single rabbit.

Table 100 shows the distribution on the same plan of the separate sexes of fleas on the separate sexes of rabbits. There were more female fleas than male, and more fleas (both male and female) were on female rabbits than on males. On the female rabbits there was evidence of a slightly greater spread with more hosts without fleas and more with high numbers than on the males. Probability diagrams were made for the two sexes of the host separately and indicated that the fit to the log normal was even closer on the males than on the females.

TABLE 100. *Fleas on rabbits as in Table 99, distributed according to the sex of parasite and host, and the level of infection in* $\times$ *3 log classes*

| | No. fleas | Fleas per host $\times$ 3 classes | | | | | | | Total rabbits | Total fleas | Fleas per host arith. mean |
		I	II	III	IV	V	VI	VII			
	Male fleas only										
On male rabbits	29	20	48	66	39	38	—	—	240	3980	16.6
On female rabbits	36	16	50	58	45	30	5	—	240	4806	20.2
On all rabbits	65	36	98	124	84	68	5	—	480	8786	18.3
	Female fleas only										
On male rabbits	23	19	43	65	45	44	1	—	240	5087	21.2
On female rabbits	28	12	46	56	59	30	9	—	240	6206	25.9
On all rabbits	51	31	89	121	104	74	10	—	480	11,293	23.5
	All fleas, both sexes										
On male rabbits	13	15	27	60	58	41	26	—	240	9067	37.8
On female rabbits	23	6	31	52	54	43	29	2	240	11,012	45.9
On all rabbits	36	21	58	112	112	84	55	2	480	20,079	41.8

These distributions are, of course, made up of twenty-four samples spread through almost a complete year, and differing greatly in the position of the mean or median value. The problem still remains as to the form of the single samples. As each of these is too small to give accurate conclusions, Table 101 was drawn up to show the number of hosts in each of the twenty-four samples which had levels of infestation either below, equal to, or above the mean log $n + 1$, or in other words the geometric mean. It will be seen that while there is some variation in individual samples, on the whole there is strong support for an equal distribution above and below this mean. Taking the twenty-four samples together there are 224 below the mean, 251 above and five equal, or in percentages 46.7, 52.3 and 1.0 per cent. There is little doubt that the general form of the distribution in the single samples is one in which the geometric mean is very close indeed to the median. This is one of the properties of a log-normal distribution, but, by itself, is not proof of exact identity. Such a relation would still be found if the distribution was either flatter, or more peaked than the normal curve; that is, some kurtosis effect might well be present, as suggested by the slightly greater spread of the fleas on female rabbits than on males.

To sum up we can say that an examination of Allan's data for fleas on rabbits gives strong support to the idea that over the whole year the frequency distribution is very close to a log normal; and further that the single fortnightly samples show on an average a distribution which has at least one character—a symmetrical distribution about the geometric mean—which is also characteristic of the log-normal distribution.

TABLE 101. *The numbers of rabbits with fleas above or below the geometric mean (from mean log $\overline{n + 1}$) for each 2-week period in 1954, showing high symmetry about the geometric mean*

Period	Below	Equal	Above	Period	Below	Equal	Above
1	11	2	7	14	11	—	9
2	10	—	10	15	9	—	11
3	6	—	14	16	12	1	7
4	7	—	13	17	11	—	9
5	11	—	9	18	9	—	11
6	6	—	14	19	9	—	11
7	10	1	9	20	8	—	12
8	11	1	8	21	8	—	12
9	12	—	8	22	9	—	11
10	8	—	12	23	10	—	10
11	10	—	10	24	9	—	11
12	8	—	12	25			
13	9	—	11	26			
				Total all periods	224	5	251
				As percentage	46.7	1.0	52.3

MITES ON RATS IN GEORGIA

Cole (1949) gives data on the frequency distribution of two species of parasitic mites, *Liponyssus bacoti* (Hurst) and *Laelaps hawaiiensis* Ewing, found on 227 rats, *Rattus norvegicus*, at Savannah, Georgia, U.S.A., in 1933.

The number of rats with 0, 1, 2, 3, etc., mites of each species is shown in Table 102 and in Fig. 89. With *L. bacoti* there were 160 rats without mites, and a total of 673 mites on the sixty-seven infested rats. With *L. hawaiiensis* there were 108 rats without mites and 1673 mites on the remaining 119 rats.

Table 102 gives the first few terms of each observed distribution together with calculations on the assumption of a log series and a negative binomial. On the basis of the log series, omitting the rats free from mites, *L. bacoti* (with $x = .9734$) gives $n_1 = 17.9$ as compared with the observed 19, and $n_2 - n_1$, 18.5 as compared with 22, both slightly low. For *L. hawaiiensis* (with $x = .9830$) n_1 is 28.4, as compared with 34, and $n_2 - n_4$ 27.3 as compared with 37—again both too low.

A negative binomial was calculated for *L. bacoti*, including the zero term ($k = .10309$), and gave estimates $n_1 = 15.9$ and $n_2 - n_4 = 18.1$, both low and poorer fits than the log series.

From the totals of the $\times$ 3 log classes in Table 103, Fig. 90 was prepared on a probability scale showing possible fits to straight lines when the mite-free rats are included, except at the highest infestations. When the mite-free rats are excluded there is a poorer fit, but a truncate log-normal is not ruled out.

TABLE 102. *Distribution of two species of mites* (Liponyssus) *on rats in Georgia, U.S.A.*

Mites per rat	*L. bacoti*	*L. hawai-iensis*	Mites per rat	*L. bacoti*	*L. hawai-iensis*	Mites per rat	*L. bacoti*	*L. hawai-iensis*
			13	— (16)	1 (25)	41	—	1
0	160	108	14	—	1	46	1	—
			15	2	—	51	—	1
1	19 (19)	34 (34)	16	—	—	52	—	1
2	11	16	17	—	1	55	—	1
3	6	14	18	1	1	58	1	—
4	5 (22)	7 (37)	19	1	1	64	1	—
5	4	7	20	—	1	65	—	1
6	4	4	21	—	2	74	—	1
7	3	3	22	—	1	89	—	1
8	2	3	27	—	1	99	—	1
9	2	2	32	—	1	100	1	—
10	—	1	33	1	—	115	1 (5)	— (8)
11	—	3	34	—	2	168	—	1
12	1	1	35	— (5)	1 (13)	329	—	1 (2)

Total rats	227	227
Rats with mites	67	119
Total mites	673	1663
Mites per rat with mites	10.05	13.95
Mites per all rats	2.96	7.33

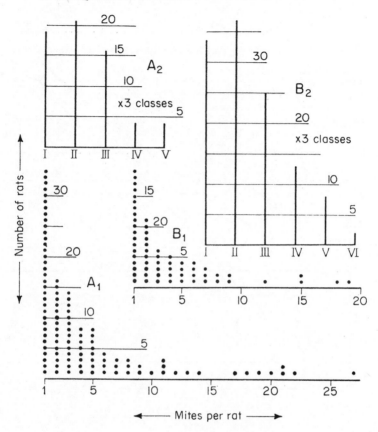

Fig. 89. Frequency distribution of rats with different numbers of two species of mites (*Liponyssus*) in Georgia, U.S.A. (Cole). A = *L. bacoti*, and B = *L. hawaiiensis*. Mites per rat shown on arithmetic and × 3 logarithmic scales.

TABLE 103. *Comparison of figures calculated from the log series and the negative binomial for the lower infestation values of mites on rats in Table 102*

Mites per rat	obs.	L. bacoti log series	neg. binom.	L. hawaiiensis obs.	log series
1	19 (19)	17.9 (17.9)	15.9 (15.9)	34 (34)	28.4 (28.4)
2	11	8.7	8.5	16	14.0
3	6	5.7	5.5	14	9.2
4	5 (22)	4.1 (18.5)	4.1 (18.1)	7 (37)	6.8 (27.3)

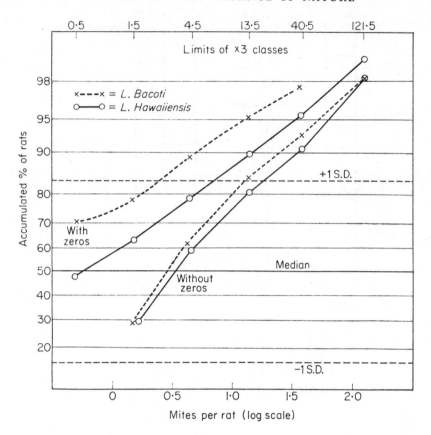

FIG. 90. Frequency distribution of rats with different numbers of mites (*Liponyssus*) plotted as accumulated percentage of total rats, on a probability scale, against number of mites per rat on a log scale. For each species of mite the results are shown with and without the inclusion of the zero term.

DISTRIBUTION OF SCABIES MITES ON MAN

In 1942 Johnson and Mellanby published the information given in Table 104 and Fig. 91 on the distribution of adult female Scabies mite, *Sarcoptis scabei* (de G.) var. *hominis* on male patients in Sheffield, England, in 1941.

From the table it will be seen that the total number of hosts was 889 and the total mites 10,699, giving an arithmetic mean of just over twelve mites per host. A calculation made on the assumption of the log series gave an estimate of 226 hosts with one mite, which is quite away from the observed 137. From the totals in the $\times$ 3 classes in the first table, Table 105 shows the accumulated total as percentages of the total number of hosts, and from this Fig. 92 has been prepared on a probability scale. It indicates a very close fit to a straight line, and we can infer that the frequency distribution of the mites

TABLE 104. *Distribution of scabies mites on man in England*

Mites per patient	No. of patients	Mites per patient	No. of patients	Mites per patient	No. of patients
1	137 (137)	17	13	41–45	9
2	123	18	11	46–50	7
3	74	19	6	51–55	5
4	68 (265)	20	9	56–60	7
5	54	21	7	61–65	6
6	56	22	3	66–70	3
7	33	23	4	71–75	2
8	30	24	8	76–80	3
9	22	25	2	93	1
10	27	26	4	116	1 (44)
11	20	27	5	166	1
12	25	28	6	254	1
13	17 (284)	29	2	280	1
14	15	30	4	320	1 (4)
15	19	31–35	14	511	1 (1)
16	9	36–40	13 (154)		

Total of 10,699 mites on 889 patients
Average (arithmetic mean) 12.04 mites per host.

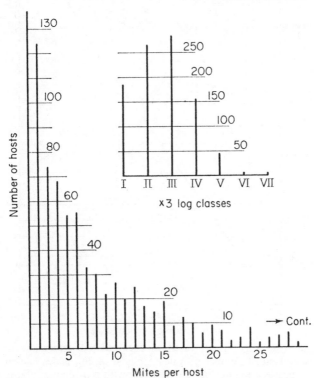

FIG. 91. Scabies mites on man. Frequency distribution of different levels of infection in arithmetic and × 3 logarithmic classes.

on the hosts is very close to a log normal with a median, on the $\log_{10}$ scale, at 0.7 ($=$ 5 mites per host) and a standard deviation of 0.53 ($=$ 3.4 arith.). It follows from this that about 68 per cent of the total hosts should have infections between the limits 0.17–(5)–1.23 on the log scale, or between 1.5–(5)–17 on the arithmetic scale, which is equivalent to 5 $\times/\div$ 3.4.

TABLE 105. *Distribution of scabies mites on man in* $\times$ *3 log classes with accumulated percentages of the total number of hosts (889)*

$\times$ 3 class	Range of class	No. of hosts	Accumulated total of hosts	As % of total
I	1	137	137	15.4
II	2–4	265	402	45.2
III	5–13	284	686	77.2
IV	14–40	154	840	94.5
V	41–121	44	884	99.4
VI	121–364	4	888	99.9
VII	365–	1	889	100

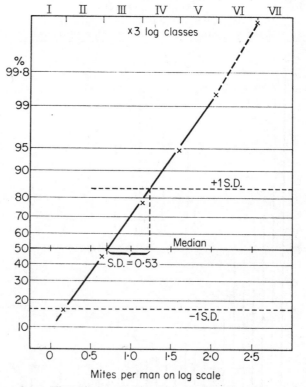

FIG. 92. Data from Fig. 91 expressed as accumulated percentage of hosts on probability scale against number of mites per host on log scale. The close fit to a straight line indicates a log-normal distribution, with a median at about 0.7 on the log scale or about five mites per man.

FILARIA WORMS IN MITES ON RATS

Having just discussed the frequency distribution of the parasitic mite, *Liponyssus bacoti*, on rats it is interesting to examine the information given by Bertram (1949) on the distribution of Filaria worms (*Litomoscoides carinii*) in the same mites. Bertram fed the mites experimentally on rats infected with Cotton Rat Filariiasis and examined microscopically 2600 mites after feeding. Of these, 1155 were negative, and 1445 gave filaria according to the distribution shown on Table 106 and Fig. 93. In all, there were 6751 filaria worms in the infected mites, giving an arithmetic mean of 4.69 per infected mite, or 2.61 for all mites examined.

Omitting the zero term, a log series to fit 6781 worms in 1445 mites gives the first four terms as shown in Table 107. The estimate is very poor. Taking all the 2600 mites and assuming the zero term correct, a negative binomial was calculated, as shown in the same table, with an even poorer fit.

Finally, Fig. 94 was drawn from the accumulated totals of the observed × 3 classes: A being with, and B without the zero term. There is a close fit to a log-normal distribution when the uninfected mites are omitted. This indicates a median on the log scale of 0.30 with a S.D. of 0.5, or on an arithmetic scale a median of 2.0 worms with a S.D. of $\times/\div$ 3.2.

TABLE 106. *Distribution of filaria worms on mites on rats*

Filaria per mite	No. of mites	Filaria per mite	No. of mites	Filaria per mite	No. of mites
		14	11	32	4
0	1155	15	5	33	2
		16	9	34	1
1	533 (533)	17	9	35	2
2	260	18	9	36	4
3	150	19	8	37	1 (103)
4	98 (513)	20	5	41	2
5	70	21	2	45	1
6	48	22	4	51	1
7	36	23	5	59	1
8	28	24	3	62	1
9	27	25	5	63	1
10	15	26	1	64	1
11	13	28	2	67	1
12	21	30	5	78	1 (10)
13	8 (266)	31	6		

Total mites, 2600; mites with filaria, 1445; total filaria, 6781; Arith. mean filaria on all mites, 2.61; on mites with filaria, 4.69

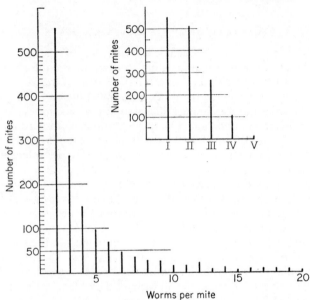

FIG. 93. Frequency distribution of filaria worms on mites (*Liponyssus bacoti*) on infected rats. In arithmetic and × 3 log classes of number of filaria per mite.

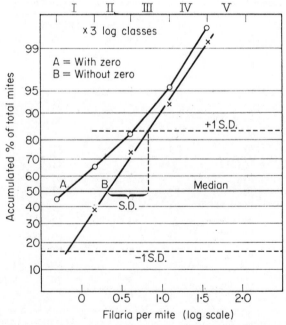

FIG. 94. Filaria in mites. Data as in Fig. 93, plotted as accumulated percentage of mites, on probability scale, against number of filaria per mite on log scale.

TABLE 107. *The earlier terms of infection, from Table 106, with calculations for comparison from the log series and the negative binomial*

	Observed	Calc. log series	Calc. neg. binomial
1	553 (553)	519 (519)	404 (404)
2	265	240	245.5
3	150	147	170.3
4	88 (513)	102.3 (489.8)	125.5 (541.3)

DISTRIBUTION OF TREMATODE CYSTS IN THE BODIES OF DRAGONFLY NYMPHS

Szidat (1931) gives figures for the number of cysts of the Trematode, *Prosthogonimus* sp. probably *pellucida*, on nymphs of the dragonfly *Cordulia aenea* collected in Germany in 1930. A total of about 975 cysts were found in 100 nymphs examined, distributed as shown in Table 108.

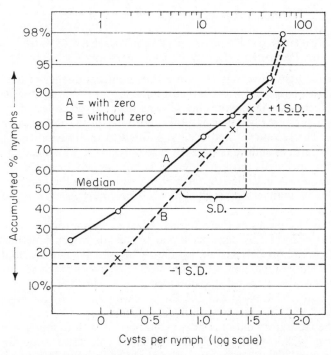

FIG. 95. Trematode cysts in dragonfly nymphs. Accumulated percentage nymphs on probability scale against number of cysts per nymph on log scale. The straight-line relation for the numbers, excluding the uninfected nymphs, suggests a log-normal distribution with a median about 0.80 on the log scale, or 6.3 cysts per nymph.

Of the 100 nymphs seventy-four were infected, so the arithmetic mean number of cysts is 9.75 approximately for all nymphs, or 13.2 for the infected hosts only. A log series, shown in the table, gives n_1 too large at 18 instead of 13, and $n_2 - n_{10}$ with a total of 32.5 instead of the observed 37.

Figure 95 shows graphically the fit of the data, on a probability scale, to a log-normal distribution. The values, omitting the uninfected nymphs, show a very close fit to a log-normal distribution with a median at 0.80 (6.3 cysts per nymph) and a S.D. of $\pm$ 0.65 ($\times/\div$ 4.5).

TABLE 108. *Distribution of Trematode cysts on nymphs of dragon-flies in Germany*

Cysts per nymph	No. of nymphs Observed	log series	Cysts per nymph	No. of nymphs
0	26	—	30–50	4
1	13	18	56	3
2–10	37	32.5	60	2
10–20	8	—	62	1
20–30	5	—	70	1
			Total	100

DISTRIBUTION OF EGGS OF A TACHINID FLY OF THE LARVAE OF PRODENIA LITURA (LEPIDOPTERA) IN EGYPT

Hafez in 1953 gave two sets of data on the numbers of eggs of the Tachinid fly, *Tachina larvarum* L., on larvae of the Cotton Worm, *Prodenia litura* (family Noctuidae) in Egypt. One set was from larvae collected in the field which were presumably under natural conditions for the choice of host by the parasites. The second lot had been submitted to the possibility of infection under laboratory conditions. In each case only larvae which had at least one worm were included, those uninfected being omitted from the count. The details are shown in Table 109 and Fig. 96, which gives the number of larvae with different numbers of eggs, and also the totals in $\times$ 3 log classes of infestation.

In the field collections there were 184 larvae with a total of 363 eggs, an arithmetic mean of 1.97 eggs per larva. In the laboratory experiments 7759 eggs were found on 2241 larvae, with an A.M. of 3.46. The distribution is of the hollow-curve type, and a log series calculated to fit each set up to n_{13} is shown in the table alongside the observed figures. The fit for the field observations is close; 103.4 instead of 105 for n_1; 64 instead of 67 for $n_2 - n_4$, and 13.8 instead of 11 for larvae with five to thirteen eggs.

For the laboratory data the fit is much poorer, with too high a calculated value for n_1 (938 instead of 842) and too low a calculated value for $n_2 - n_4$ (818 instead of 896). This excess in the observed number of larvae with two, three or four eggs and a shortage of those with one egg only is what might be expected in the more enclosed conditions of the laboratory, where the flies would not have the same freedom to roam as in the field.

TABLE 109. *Distribution of eggs of parasitic Tachinids (Diptera) in larvae of* Prodenia litura *(Noctuidae) in Egypt, from field samples and laboratory infections*

	Field sample			Laboratory sample			
	obs.	log series		obs.	log series		obs.
1	105 (105)	103.5 (103.5)	1	843 (843)	938 (938)	15	18
2	41	37.0	2	423	415	16	7
3	14	17.6	3	268	243	17	5
4	12 (67)	9.4 (64.0)	4	205 (896)	160 (818)	18	4
5	2	5.4	5	110	112.5	19	6
6	5	3.2	6	102	82.3	20	3
7	1	2.0	7	62	62.1	21	3
8	2	1.3	8	42	47.6	22	2
9	1	0.79	9	37	37.1	23	2
10	—	0.51	10	36	29.3	24	1
11	—	0.33	11	30	23.4	25	1
12	—	0.20	12	13	18.9	28	1
13	— (11)	0.14 (13.8)	13	9 (441)	15.3 (428)	29	2
14	1	—	14	14	—	30	1

Total larvae		184	Total larvae		2241
Total eggs		363	Total eggs		7759
A.M. eggs per larva		1.97	A.M. eggs per larva		3.46

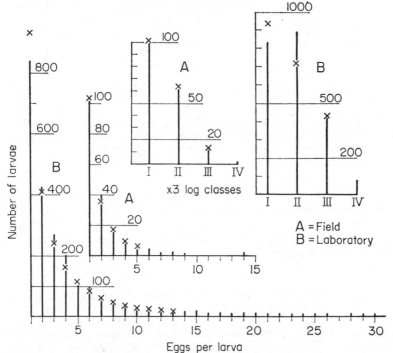

FIG. 96. Eggs of parasitic Tachinidae (Diptera) in larvae of the moth *Prodenia litura* in Egypt. A = field samples and B = laboratory infections; showing frequency distribution with number of eggs on arithmetic and × 3 log scales.

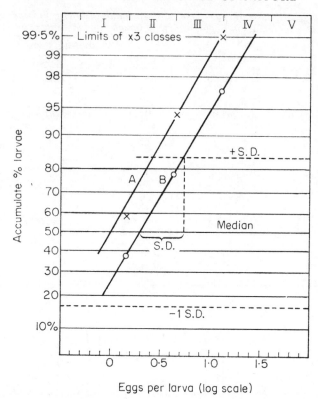

Eggs per larva (log scale)

FIG. 97. Eggs of parasitic Tachinidae (Diptera) on larvae of the moth, *Prodenia litura*, in Egypt. Data as in Fig. 96, showing relation between accumulated percentage larvae on probability scale and number of eggs per larva on log scale. Both field and laboratory data support a log-normal distribution.

Figure 97 shows the two sets of data plotted as accumulated percentages against a probability scale. Both give a good fit to a log-normal distribution with a higher median, 0.25, in the laboratory sample than in the field sample, which is 0.00 or 1 egg per larva. In each case the standard deviation, on the log scale, is approximately ± 0.43.

DISTRIBUTION OF EGGS OF AN ICHNEUMON FLY IN LARVAE OF EPHESTIA UNDER LABORATORY CONDITIONS

In contrast to the distribution of various parasites on their hosts under natural conditions it is of interest to consider data published in 1943 by F. J. Simmons, who subjected larvae of *Ephestia kuhniella* Zeller (Lepidoptera, Pyraustidae) to attack, in laboratory cages, by various numbers of females of the Ichneumon fly *Nemeritis canescens* Grav.

TABLE 110. *Distribution of eggs of an Ichneumon fly in larvae of Ephestia (Lepidoptera) under laboratory conditions with different ratios of female parasites to hosts*

	I	II	III	IV	V	VI	VII
Parasite to host ratio	1/200	1/100	1/50	1/25	2/25	5/25	10/25
Eggs per host 0	1351	493	153	21	2		
1	486	339	183	77	14		
2	66	89	79	52	23	} 29	0
3	2	22	33	41	25		
4	—	9	13	23	19		
5	—	3	7	13	15		
6	—	—	2	9	5		
7	—	1	—	1	4	} 74	4
8	—	—	1	2	4		
9	—	1	1	—	2		
10	—	—	—	2	5		
11	—	—	—	—	1	} 76	20
12	—	—	—	—	3		
14	—	—	—	—	1		
15	—	—	—	1	1	} 42	16
18	—	—	—	—	1		
20–24	—	—	—	—	—	10	20
25–30	—	—	—	—	—	1	18
30–34	—	—	—	—	—	2	14
35–39	—	—	—	—	—	—	10
40–44	—	—	—	—	—	1	5
					one at		
45–49	—	—	—	—	45	—	1
50–54	—	—	—	—	—	—	4
54–59	—	—	—	—	—	—	1
Total larvae with eggs	554	464	319	221	124	235	113
Total larvae	1905	957	472	242	126	235	113
Total eggs	624	650	556	573	583	(2750)	(2799)
Aver. eggs per larva	0.325	0.679	1.178	2.36	4.63	11.70	24.77

Seven sets of experiments were made in which the ratio of Ichneumons to larvae were in the ratios 1/200, 1/100, 1/50, 1/25, 2/25, 5/25 and 10/25. The resulting number of larvae with different numbers of eggs in each set is shown in Table 110, and in Fig. 98 for the first five sets only. As the general form of the distributions seemed to resemble a Poisson type of series, which would be the result of a random distribution of eggs by each female, values for the first few terms of such series were calculated, using the total number of larvae and the arithmetic mean eggs per larva, for the first five sets with the results shown in Table 111 and also superimposed on Fig. 98.

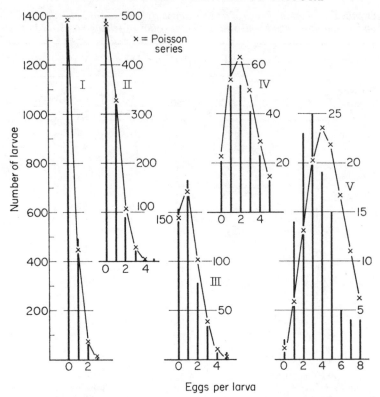

Eggs per larva

FIG. 98. Frequency distribution of eggs of Ichneumon flies in larvae of the moth Ephestia, under laboratory conditions with different proportions of female Ichneumons to Ephestia larvae. Results for the Poisson series for comparison show a good fit for low infection, but an increasingly poor fit for higher infection levels.

TABLE 111. *Comparison of results in Table 110 with lower values obtained from a Poisson series, indicating a random distribution*

Eggs per larva	I		II		III		IV		V	
	Obs.	Pois-son	Obs.	Pois-son	Obs.	Pois-son	Obs.	Pois-son	Obs.	Pois-son
0	1351	1377	493	485	153	145	21	22.7	2	1.2
1	486	447	339	329	183	171	77	53.7	14	5.7
2	66	73	89	108	79	101	52	63.5	23	13.2
3	2	8	22	25.3	33	39.6	41	50.1	25	20.4
4	—	0.6	9	4.3	13	11.7	23	29.7	19	23.6
5	—	0.04	3	0.6	7	2.8	13	14.0	15	21.9
Over five	0	0	2	0.2	4	1	15	8.3	28	23.2

It will be seen that the Poisson series gives very close estimations of the observed data in the cases where the ratio of parasite to host was low and the average number of eggs per larva less than one. In Set III, where there was one parasite to fifty hosts and a resulting average infection of 1.18, there is a slight excess of observed over calculated in the larvae with one egg, and the reverse in those with two eggs. This condition is more marked in Set IV with one female to twenty-five larvae and an average infection of 2.36 eggs per larva. Here the observed n_1 is considerably higher than the calculated and the excess is balanced by deficiencies in n_2, n_3 and n_4. In Set V with two ichneumons per twenty-five larvae and an average infection of 4.6 eggs per larva, the departure is still more marked, with a large excess of observed cases in n_1, n_2 and n_3 balanced by deficits in all Classes n_4 to n_8. The numbers remaining with more than five eggs, as shown in the table, indicate that in Sets III, IV and V there are always rather more remaining in the observed than in the calculated. The number of larvae without any eggs is consistently very close to the estimates at all levels of infection.

So we see that, under the artificial conditions of the laboratory, a low ratio of parasite females to host larvae results in a distribution of eggs which is almost random. As the ratio of parasite to host increases, the distribution becomes less and less random in the direction of the observed figures being larger than the calculated with small and with high numbers of eggs per larva, while they are smaller when the infestation is about the average.

In experimental Sets VI and VII the classes of infection were grouped in fives, so the closeness of fit for any level is not so easy to check; it is, however, interesting to note that in both sets the median larva infection level is, as near as can be estimated, very close to the average number of eggs per larva. In Set VI the average is 11.7 and the median approximately 11; in Set VII the average is 24.8 and the median about 23. In nearly all the distributions previously examined the median has been near the geometric mean level of infection and usually well below the arithmetic mean.

HIPPOBOSCID FLIES ON QUAIL IN CALIFORNIA

In 1958, Tarshis published an account of the occurrence of two species of Hippoboscid flies on two sub-species of quail in California. The host birds were both *californica* and *Lophortyx c. brunescens*. The flies belonged to two different genera, *Stilbometopa impressa* (Bigot) and *Lynchia hirsuta* Ferris.

In all, 3313 birds were examined; this, however, included a few which were recaptured and so were counted twice. *S. impressa* was found on 1828 birds and *L. hirsuta* on 1016. The details of the presence or absence of flies, together with the number of flies of each species was given individually for 1665 birds and is summarized in Table 112.

First it will be seen that the average incidence on birds with flies is very low—less than two—and no bird had more than eight flies. The frequency distribution is the familiar hollow curve (Fig. 99). Neglecting the zero term, it can

TABLE 112. *Distribution of two species of Hippoboscid flies on quail in California*

Flies per bird	First captures		Birds caught again		Total birds obs.	cal. log series
	Stilbometopa impressa					
	impressa only	both species	*impressa* only	both species		
1	243	48	43	4	388	354
2	97	14	12	—	123	102
3	32	11	2	—	45	40.7
4	19	4	1	—	24	17.9
5	4	—	—	—	4	4
6	1	1	—	—	2	
7	2	—	—	—	2	
Total birds	398	78	58	4	538	
Total flies	649	127	75	4	861	
Av. flies per bird	1.63	1.63	1.24	1.00	1.60	

Number of birds without *S. impressa* = 1127

	Lynchia hirsuta					
	hirsuta only	both species	*hirsuta* only	both species		
1	120	46	3	3	172	182
2	44	22	1	—	67	58.3
3	19	7	1	1	28	25.1
4	6	1	—	—	7	12.0
5	10	1	—	—	11	
6	2	1	—	—	3	
7	1	—	—	—	1	
8	1	—	—	—	1	
Total birds	203	78	5	4	290	
Total flies	366	126	8	6	506	
Av. flies per bird	1.80	1.62	1.60	1.50	1.74	

Number of birds without *L. hirsuta* = 1375.

Total number of birds = 1665

be shown that if the distribution were in the form of a log series the number of birds with only one *S. impressa* should be 354 (observed 338); and with only one *L. hirsuta*, there should be 182 (observed 172). In each case the calculated value is about 5 per cent too high. The total number of classes is rather small for reliable conclusions.

It will be noticed also that the average number of *S. impressa* per bird is 1.63 when it is the only species present, and also when *L. hirsuta* is present in addition. In the case of *L. hirsuta*, however, the average is 1.80 when alone and 1.62 in the presence of *S. impressa*. The difference is small, but suggests the possibility that the presence of *L. hirsuta* does not interfere with *S. impressa*, but that the presence of *S. impressa* may interfere with *L. hirsuta*.

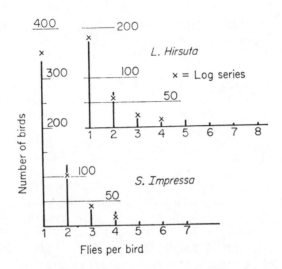

FIG. 99. Frequency distribution of two species of Hippoboscid flies on quail in California, together with numbers estimated from the logarithmic series which show a very close fit to the observed values.

HIPPOBOSCID FLIES ON BIRDS IN FAIR ISLE, SCOTLAND

For comparison with the Californian results just quoted, Table 113 shows some data obtained in 1954 by Corbet (1956) on Fair Isle, about 80 miles N.N.E. of the most northerly point of Scotland.

The Hippoboscid was *Ornithomyia fringellina* Curtis (said to be possibly the same as *O. lagopodis* Sharp) and were found chiefly on five hosts, Starling (*Sturnus vulgaris*), Wheatear (*Oenanthe oenanthe*), Rock Pipit (*Anthus spinoletta*), Meadow Pipit (*Anthus pratensis*), and Twite (*Cardualis flavirostris*). Other species were also found infested but more rarely.

It will be seen from the table that out of 727 birds examined 380 had no flies; and the remaining 347 birds carried a total of 691 flies, distributed in the familiar hollow curve, with 164 birds with only a single fly on each. The average for all infested birds was 1.99 flies per bird, ranging from 1.63–2.15 on the five different hosts. The maximum infection was ten flies on one wheatear.

TABLE 113. *Distribution of Hippoboscid flies on different species of birds on Fair Isle, north of Scotland*

Host	Starling	Wheat-ear	Rock pipit	Meadow pipit	Twite	Total obs.	Total log ser.
Total birds	117	216	217	116	61	727	
Flies per bird							
0	47	107	104	75	47	380	
1	30	48	56	23	7	164	193.5
2	17	31	33	13	5	99	69.7
3	11	16	17	3	2	49	30.4
4	3	7	4	1	—	15	18.1
5	5	2	2	1	—	10	10.4
6	2	1	—	—	—	3	6.2
7	1	2	1	—	—	4	3.9
8	1	1	—	—	—	2	2.4
9	—	—	—	—	—	—	1.5
10	—	1	—	—	—	1	1.0
					Over 10	0	9.8
Total birds with flies	70	109	113	41	14	347	
Total flies	161	234	206	67	23	691	
Av. flies per bird with flies	2.3	2.15	1.82	1.63	1.64	1.99	
Av. flies all birds	1.37	1.09	0.95	0.58	0.38	0.95	

× 3 log classes	Observed	Log series
I	164	193.5
II	163	118.2
III	20	26.9

Figure 100 shows the frequency distribution of birds with no flies, and with 1, 2, 3, etc., and also the distribution in × 3 log classes. On the table and figure is also given the frequency distribution calculated from a log series with 691 flies on 347 hosts. It will be seen that the calculated n_1 is distinctly too high, and n_2 and n_3 too small. The log series also requires that ten birds should have had more than ten flies each. Calculations for the n_1 value for each host separately gave a too-high value in every case. The fit is not so good as in the Californian data.

The absence of high infestations may, however, be due to a new biological effect, as Corbet says that the highest mortality among the flies is probably due to the preening of the birds.

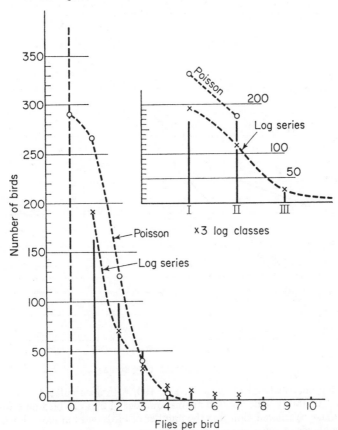

FIG. 100. Frequency distribution of Hippoboscid flies on birds in Fair Isle, Scotland, with arithmetic and × 3 classes of abundance. Also numbers calculated on the basis of the Poisson series and logarithmic series for comparison.

Figure 101 shows a test of the fit of the observed data to a log-normal distribution, by means of a probability graph. Two tests were made, one including and one excluding the uninfested birds. As in the case of the Californian data, the best fit is obtained by excluding the uninfested birds. This suggests the possibility of a log normal with a median just above 0.20, and a S.D. of about 0.28 on the log_{10} scale, or 1.6 $\times / \div$ 1.95 on the arithmetic scale.

A final test was made with a Poisson series to see if the distribution of the flies was entirely random. The series calculated from 727 birds with an average of 0.95 flies per bird gave the following values (with the observed numbers in brackets):

No flies 281 (380); one fly, 267 (164); two flies, 127 (99); three flies, 40 (49) and four flies, 9.5 (15). There is thus no evidence of a purely random distribution. The flies are aggregated on too few birds for this to have occurred.

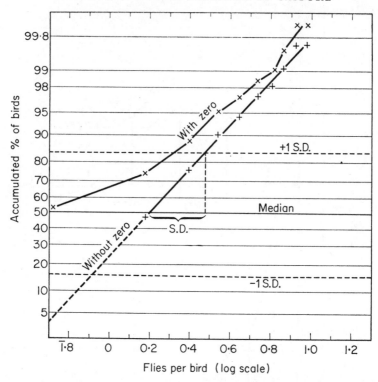

FIG. 101. Hippoboscids on birds, as in Fig. 100, with accumulated percentage of birds on a probability scale, plotted against number of flies per bird on a log scale. The percentage calculated, omitting the number of birds without flies, shows a very close resemblance to a log-normal distribution.

COPEPODS (CRUSTACEA) PARASITIC ON MUSSELS IN BRITAIN

In 1954, G. D. Waugh published an account of the occurrence of the parasitic Copepod, *Mytilicola intestinalis* (Steuer) in the Mussel *Mytilus edulis* on the English coast. Through the kindness of this author and of B. T. Hepper, who continued the investigation, I received a copy of their original data with the number of parasites found in each mussel in 128 samples, usually of twenty mussels, in various localities on the coasts of England and Wales. These samples contained a total of 2236 mussels infected by 11,092 parasites, giving an overall arithmetic mean of 4.96 parasites per host. There was, however, considerable variation in the level of infection in different localities.

From the data it was easy to find for each sample the total number of parasites and the number of mussels infected. The relation between the two is shown diagrammatically on Fig. 102 with the number of parasites per sample on a logarithmic scale. The solid line A shows the distribution expected if each parasite avoided as far as possible any already infected host, that is to say complete dispersal, so that the first twenty parasites per sample would infect every mussel. The dotted line B shows (from a binomial calculation) the expected relation between total number of parasites and the number of infected hosts, out of twenty, on the assumption of a purely random distribution.

It will be seen that although a few of the samples are on this random line, the great majority show too small a number of mussels infected for the number of parasites present. Thus in the sample marked with an asterisk in the figure there were 174 parasites in twenty mussels, and yet three hosts were still free of infection. It can be shown that a random distribution of only

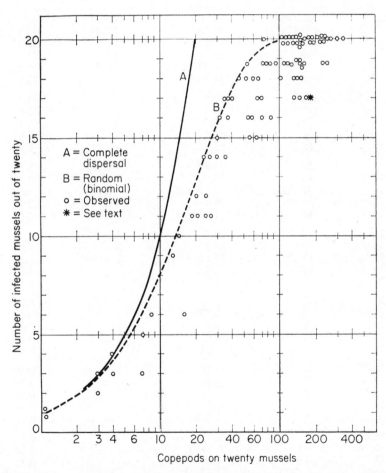

FIG. 102. Relation between number of copepod parasites in samples of twenty mussels from various localities in Britain and the number of mussels per sample free from infection. A = the line for complete dispersal and B (dotted) for a random distribution of the parasites.

TABLE 114. *Frequency distribution of parasitic copepods in mussels from various localities round the coast of England, and also from samples from Poole Harbour (Dorset) only. For comparison with the observed numbers of mussels calculations based on a negative binomial have been added*

	All samples		Poole Harbour only	
No. of samples	126		13	
Total mussels	2236		250	
Total copepods	11,092		1226	
Parasites per host, arithmetic mean	4.96		4.90	

	Observed	Negative binomial	Observed	Negative binomial
Zero	460 (460)	460* (460*)	47 (47)	47* (47*)
1	271 (271)	318.5 (318.5)	37 (37)	35.5 (35.5)
2	239	247.1	19	28.5
3	193	198.6	22	23.3
4	178 (610)	162.5 (608.2)	22 (63)	19.1 (70.9)
5	148	134.3	14	15.9
6	124	110.9	16	13.2
7	111	93.3	16	11.0
8	77	78.2	9	9.16
9	66	65.8	7	7.66
10	58	55.5	8	6.41
11	51	46.8	7	5.37
12	40	39.6	4	4.51
13	39 (713)	33.5 (657.8)	5 (86)	3.70 (76.9)
14	35	28.4	4	3.18
15	14	24.1	2	2.67
16	19	20.5	1	2.24
17	19	17.4	2	1.89
18	17	14.8	1	1.59
19	14	12.6	2	1.33
20	6	10.7	—	1.12
21	17	9.12	1	0.94
22	5	7.77	1	0.80
23	8	6.63	1	0.67
24	4	5.65	—	0.56
25	3	4.82	—	0.48
26	3	4.11	1	0.40
27	6	3.51	—	0.34
28	2	2.99	—	0.28
29	—	2.56	—	0.24
30	1	2.18	1	0.20
31	3	1.87	—	0.17
32	—	1.59	—	0.14
33	3	1.36	—	0.12
14–40	(179)	(188)	(17)	(19.8)
over 40	45, 53 (2)	(3.62)	0	

* By assumption.

forty parasites would leave this same number free. Thus there is evidence of a departure from random in the direction, not of greater dispersal, but of considerable aggregation. In other words a mussel that has already received one parasite is more likely to get a second or a third than one which is not already infected. This could be due to a direct selection by the parasites of an already

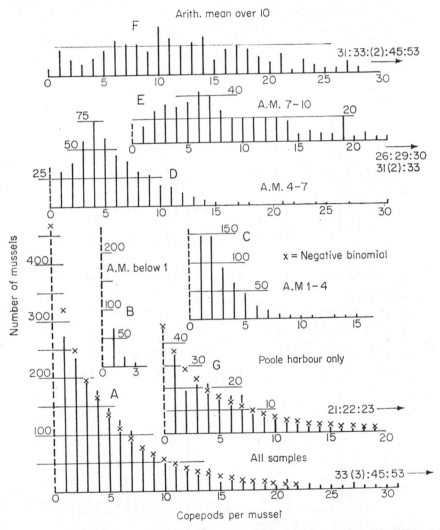

FIG. 103. Frequency distribution (on arithmetic scale) of mussels with different numbers of parasitic copepods in samples including 2236 mussels from various parts of the S.E. coast of England. A = all samples; B = Poole Harbour only; C–F = various levels of arithmetic mean infestation in samples of twenty mussels. xx = calculations based on the assumption of a negative binomial distribution.

infected host, or to some greater attraction or repulsion by certain mussels; but is more likely to be due to variation in position of hosts in the mussel bed leading to more easy or more frequent access.

Table 114 shows the number of mussels with different numbers of parasites, first in all samples from all localities together, and secondly in the thirteen samples taken from Poole Harbour in Dorset. The frequency distributions are shown diagrammatically in Fig. 103, using an arithmetic scale of parasites per host, and in Fig. 104, using geometric and square-root transformations.

For the total of all samples there were 11,092 copepods in 2236 mussels examined, of which 430 were free from infection, giving an overall arithmetic mean of 4.93 parasites per host. The frequency distribution is of the type with

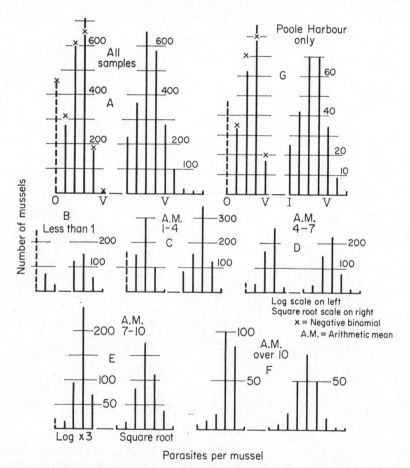

FIG. 104. Data for frequency distribution of mussels with different number of copepod parasites as in Fig. 103, but grouped in × 3 classes (on left) and in square-root classes (on right). Lettering as in previous figure.

more zeros than ones, more with one than with two, etc., but is definitely not of the log-series form. When the numbers are grouped in × 3 classes the resulting distribution is overcorrected for symmetry, but when square-root classes are used the result is almost symmetrical (Table 115 and Fig. 104).

TABLE 115. *Mussels with parasitic copepods. Data as in Table 114 arranged in geometric × 3 classes, and in square-root classes*

	All samples	below 1	Samples with arithmetic means 1–4	4–7	7–10	over 10	Poole Harbour
Geometric classes × 3							
Zeros	460	243	163	35	16	3	47
I	271	69	151	28	14	9	37
II	610	26	301	174	93	16	63
III	714	—	100	263	249	102	86
IV	179	—	2	23	69	85	17
V	2	—	—	—	—	2	—
Square-root classes, divisions at integers							
I	230.0	121.5	81.5	17.5	8.0	1.5	23.5
II	365.5	156.0	157.0	31.5	15.0	6.0	42.0
III	656.5	59.5	345.0	150.5	84.0	17.5	70.5
IV	582.0	1.0	123.5	230.5	175.5	51.5	69.5
V	279.5	—	10.0	79.0	114.5	76.0	34.0
VI	101.0	—	—	11.5	38.0	51.5	8.5
VII	19.5	—	—	2.5	6.0	11.0	2.0
VIII	1.0	—	—	—	—	1.0	—
IX	1.0	—	—	—	—	1.0	—

These two transformations also are shown against probability scales in Fig. 105 (as broken lines) and the closeness of the square-root relation to a straight line is seen. To test the possibility of a negative binomial distribution a calculation was made to fit 11,025 parasites in 2236 hosts with a zero term of 460. This, with an exponential of 0.8054, gave the series shown alongside the observed values in Table 114 and in Figs. 103 and 104. This gives a close fit to the observed data, except that the calculated values are too high for hosts with one and two parasites, and generally a little too low between four and fourteen.

The thirteen samples from Poole Harbour (Table 114 and Figs. 103 and 104) give in all three tests almost identical results although containing only about one-tenth of the population. The arithmetic mean parasites per host is 4.9 instead of 4.93. The form of distribution is quite similar on the arithmetic scale, the × 3 log scale and on the square-root transformation, and when transferred to log and square root × probability diagrams the Poole Harbour and the total sample lines are almost indistinguishable. Finally when a negative binomial is calculated the exponential is 0.8944, slightly larger than that for the whole sample.

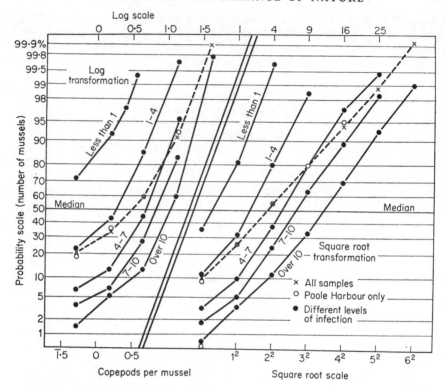

Fig. 105. Accumulated percentage frequencies, on a probability scale, of mussels with different numbers of copepod parasites, plotted on log and square-root scales. Separate lines in each half of the diagram give the total sample, the samples of Poole Harbour, and samples with different arithmetic mean levels of infestation. A straight line would indicate a log-normal or a square-root normal distribution.

Since the arithmetic mean number of parasites per host varies in the different samples from 0.05 to 16.75, and the actual number in a single host from 0 to 53, the data lend themselves to a study of the frequency distribution of the parasites at different levels of infection. Table 116 shows the number of hosts with different numbers of parasites in subdivisions of the original data: first in all samples with a mean of less than one parasite per host; then three divisions with means of 1–3.95, 4–6.95, and 7–9.95; and lastly the heavily infested samples with arithmetic means of over ten parasites per host. The results are shown diagrammatically in Figs. 103 and 105.

It will be seen that, on the arithmetic scale the number of hosts with 0, 1, 2, 3, etc., parasites falls steadily from the zero to the larger numbers, both in the total population and at all infection levels up to an average of four parasites per host. Above this the peak comes at a higher level: at about 4 in the 4–7 level of infection, at 6 for the 7–10 level, and at about 10 for the over

TABLE 116. *Frequency distribution of parasitic copepods in mussels arranged in groups according to the average density of infection in samples of twenty mussels*

	below 1	Samples with arithmetic means			over 10
		1–4	4–7	7–10	
No. of samples	19	40	32	24	11
Total mussels	338	717	523	441	217
Total copepods	133	1628	2996	3584	2710
Parasites per host	0.39	2.26	5.67	7.91	12.49
Parasites per host 0	243	163	35	16	3
1	69	151	28	14	9
2	16	148	42	27	6
3	8	90	57	34	4
4	2	63	75	32	6
5	—	44	60	35	9
6	—	21	46	45	12
7	—	16	42	41	12
8	—	8	31	28	10
9	—	6	28	21	11
10	—	3	19	22	14
11	—	1	18	19	13
12	—	—	11	19	10
13	—	1	8	19	11
14	—	1	5	16	13
15	—	1	3	5	5
16	—	—	2	8	9
17	—	—	3	6	10
18	—	—	1	7	9
19	—	—	—	8	6
20	—	—	—	2	4
21	—	—	3	6	8
22	—	—	2	2	1
23	—	—	—	3	5
24	—	—	1	—	3
25	—	—	1	—	2
26	—	—	—	1	2
27	—	—	1	1	4
28	—	—	—	—	2
29	—	—	—	—	—
30	—	—	—	1	—
31	—	—	1	2	—
33	—	—	—	1	2
45	—	—	—	—	1
53	—	—	—	—	1

10 level of infection. In the latter group there are no zeros. When the scale is transformed, the square root gives more symmetrical results with the peak in Class III for all samples, and in III–IV for Poole Harbour.

For the different levels of infection, graphs against a probability scale are shown in Fig. 105, for both square root and $\times$ 3 transformations. In every case the square root gives a closer fit to a straight-line relation than the log transformation.

One can conclude that the negative binomial and the root normal represent the observed facts better than either the log normal or the log series. The very close similarity between the frequency distribution for Poole Harbour only, and the much more diverse conditions of all the samples together, is quite remarkable.

COPEPOD AND BRANCHIURAN PARASITES IN FISHES IN EAST AFRICA

An interesting comparison with the last section, dealing with Copepods in sedentary mussels, is found in some work by Fryer (1956, 1959, 1961) on the occurrence of Copepods and Branchiura as external parasites on fresh-water fishes in East African lakes. In these the parasite has to attach itself to a mobile host.

Mr Fryer has kindly given me some of his original figures for the number of parasites per host in eight different host-parasite relations, which are summarized in Table 117. Three of the parasites are Copepods, *Lernea hardingi*,

TABLE 117. *Frequency distribution of fresh-water fishes in East African lakes with different numbers of parasitic copepods and branchiura*

Number of hosts	Arith. parasit. mean P. per H.	\multicolumn{11}{l}{Number of parasites per host}
		0	1	2	3	4	5	6	7	8	9	10	over 10 at

A. *L. hardingi* on *C. mabusi* in Lake Bangweulu
| 477 | 150 | 0.31 | 376 | 73 | 19 | 4 | 2 | — | 1 | 1 | 1 | — | — | |

B. *L. bagri* on *B. meridionalis* in Lake Nyasa
| 560 | 1183 | 2.11 | 227 | 116 | 70 | 37 | 35 | 27 | 10 | 4 | 5 | 6 | 4 | 11 (3), 12 (6), 13 (4), 14, 18, 21, 24, 30, 55. |

C. *L. cyprinacea* on *T. esculenta* in Lake Victoria
| 285 | 48 | 0.17 | 272 | 4 | 5 | 1 | 1 | — | — | — | 1 | — | 18. | |

D. *L. cyprinacea* on *T. variabilis* in Lake Victoria
| 604 | 212 | 0.35 | 532 | 27 | 15 | 7 | 9 | 6 | 3 | — | 2 | 1 | 1 | 15. |

E. *A. africanus* on *T. variabilis* in Lake Victoria
| 650 | 412 | 0.63 | 458 | 95 | 55 | 15 | 16 | 2 | 5 | 2 | — | — | — | 12, 32. |

F. *A. africanus* on *T. esculentus* in Lake Victoria
| 342 | 1086 | 3.18 | 154 | 70 | 25 | 22 | 6 | 8 | 4 | 8 | 3 | 1 | — | 11 (4), 12 (7), 13, 14 (3), 15 (2), 16, 17 (3), 18 (4), 19 (3), 20 (3), 21 (2), 22 (2), 24 (2), 28 (2), 30, 34. |

G. *D. ranarum* on *T. variabilis* in Lake Victoria
| 650 | 178 | 0.27 | 531 | 84 | 24 | 5 | 8 | 1 | — | — | — | — | — | |

H. *D. ranarum* on *T. esculentus* in Lake Victoria
| 342 | 127 | 0.37 | 283 | 39 | 6 | 9 | 2 | — | — | 1 | 1 | — | — | 26. |

L. bagri and *L. cyprinacea*; and the other two Branchiura of the family Argulidae, *Argulus africanus* and *Dolops ranarum*. The host fishes here discussed are *Chrysichthys mabusi* in Lake Bangweulu, *Bagrus meridionalis* in Lake Nyasa, and *Tilapia esculenta* and *T. variabilis* in Lake Victoria.

Table 117 gives the total number of hosts and of parasites for each group, and the frequency distribution of hosts with different numbers of parasites. The percentage of hosts free from infection range from 41 per cent in Group B to 95 per cent in Group C. The highest number on a single host in these data was 55, but higher numbers were found in other samples. The frequency distributions are shown diagrammatically in Fig. 106.

Fryer studied the question of randomness of dispersion by comparison with a Poisson series, calculated to fit the correct number of hosts and parasites. His calculated values for n_0 and n_1 are given in Table 118, and it will be seen that in every case the observed n_0 (hosts free from parasites) is above that

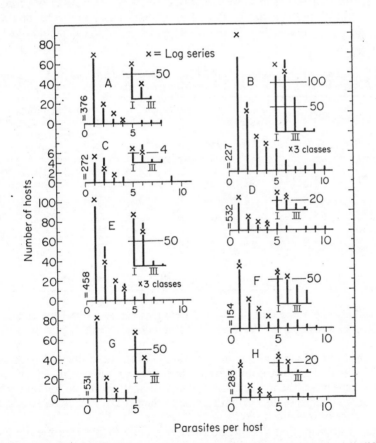

FIG. 106. Frequency distribution of East African fresh-water fishes with different numbers of parasitic copepods and branchiura. (For key, see Table 117.)

calculated for a random distribution. In Group F, with a high average level of infection, the observed n_0 is over ten times the calculated number. For n_1 there are fewer observed than calculated in all cases except Group F, where the calculated n_0 was very low. It is thus established that the parasites are distributed on a smaller, and sometimes much smaller, number of hosts than would be expected by a purely random distribution.

Table 118 also shows the parasite–host distribution in each of the eight groups when the numbers of parasites per host are classified on a × 3 scale. These results are also shown in the smaller diagrams in Fig. 106.

Calculations were also made to see the fit of the log series to the data, omitting the zero term. Table 118 shows, for comparison with the × 3 classes, the calculated log series values for n_1 and $n_2 - n_4$. These values (including n_2, n_3 and n_4 separately) are shown as crosses in Fig. 106. It will be seen that there is in general a close fit between the observed and the calculated values, and in some cases the fit is remarkably close. For example in Group A, n_1 observed is 73 and calculated 70, and n_2 is observed 21 and calculated 18.7. In Group G, n_1 calculated is 84 and observed 83, while n_2 calculated is 21 and observed 22.2.

TABLE 118. *East African fishes infected with copepods and branchiura, in × 3 classes of infection level, and with Poisson and log-series calculations for comparison*

			× 3 Class				Poisson*		Log series	
	0	I	II	III	IV	V	n_0	n_1	n_1	n_2-n_4
A	376	73	25	3	—	—	349	109	70	30.3
B	227	116	142	69	5	1	68	143	138	120.8
C	282	4	7	1	1	—	241	40	5.3	4.7
D	532	27	31	13	1	—	428	149	34.3	28.5
E	458	95	86	10	1	—	345	218	103.8	69.0
F	154	70	53	36	29	—	14	45	62.3	60.5
G	531	84	34	1	—	—	495	135	83.1	33.2
H	283	39	17	2	1	—	236	87	32.0	21.3

* From Fryer.

It will also be noted that the small differences between observed and calculated values show no regular bias, being sometimes positive and sometimes negative.

It is not possible to make an accurate study of the fit of the data to the log-normal distribution owing to the high proportion of the hosts which come into the zero class, theoretically containing all hosts with fractional numbers of parasites, but actually including many which have not come within the same chance of infection.

Thus it appears that once again there is a close fit between the observed distribution of fishes with different numbers of parasitic Copepods and Branchiura and the logarithmic series, over a range of arithmetic mean infection from 0.17 to 3.18 parasites per host. If the uninfected fishes are excluded, the arithmetic mean number of parasites per infected host ranges from 1.5 in Group G to 5.78 in Group F.

DISTRIBUTION OF EGGS OF PARASITIC NEMATODES IN THE FAECES OF SHEEP IN SCOTLAND

In 1952, Hunter and Quenouille made a statistical study of the interpretation of the number of eggs of parasitic nematodes found in samples taken from the faeces of sheep in different parts of Scotland. The technique was to take a sample of faeces from the rectum of each sheep, from which two grammes were later weighed out and diluted with 60 ml of saturated saline. From this suspension, immediately after mixing, a sample was taken with a pipette, and placed in a specially graduated slide (McMaster) so that under the microscope the number of eggs in 0.15 ml of suspension could be counted. Thus the final count involves the level of infection in the particular sample from the sheep and also the difference between samples from the same suspension.

A preliminary examination of a large number of samples from single sheep led to the conclusion that in this case the frequency distribution of the number of eggs per count followed closely to a Poisson series. The samples from a large number of sheep, on the other hand, were close to a negative binomial.

Hunter and Quenouille suggest that if the mean infestation in different sheep follows a logarithmic series, and the different counts from a single sheep follow a Poisson distribution, then the combined result, as shown by counts from many sheep, will be a negative binomial, as found.

The actual frequencies found are only given for one set of ninety sheep from a heather hill in Perthshire. These are shown in Table 119 and Fig. 107, together with a negative binomial calculated from $p = 11.2$ and $k = 0.6$. The values for these constants are also given for sheep from thirteen other localities, the values of p ranging from 0.89–16.9 and the values of k from 0.42–1.45.

On the question of the interpretation of these results Hunter and Quenouille say: "Generally speaking, k is an index of dispersion; a high value of k indicating a near random, or Poisson, distribution, a low value indicating a highly variable population. On the other hand p is more sensitive to changes in the mean count and depends on the dilution used in the counting. . . . In fact, pk is equal to the mean of the distribution, but p indicates changes in the mean due to a *general* increase or decrease rather than due to the introduction of a few more extreme values."

On the interpretation of the particular data in hand they say: "There is large variation in p due chiefly to different levels of egg output, but k remains relatively constant for different breeds, places and months. There is however a strong suggestion that 'heather hill' gives lower values of k than 'grass' or 'pasture', the overall values of k being 0.54 ± 0.07 and 0.78 ± 0.07 respectively. This may be expected from our knowledge of the grazing habits of sheep. The sheep in grass graze more evenly than those in heather and are

TABLE 119. *Distribution of eggs of parasitic nematodes in sheep in Scotland together with numbers calculated on the basis of the negative binomial*

Eggs per count	Observed frequency	Calculated negative binomial	Eggs per count	Observed frequency	Calculated negative binomial
0	20	22.61	18	1	—
1	12	12.21	21	1	—
2	14	8.79	22	2	—
3	7	6.86	24	1	—
4	3 }9	5.56 }10.16	26	1	—
5	6	4.60	27	1	—
6	3	3.86	29	1	—
7	3 }11	3.28 }12.35	31	1	—
8	2	2.80	38	1	—
9	3	2.41	40	1	—
10	2	—	66	1	—
11	2	—	10–66	17	17.02
15	1	—			

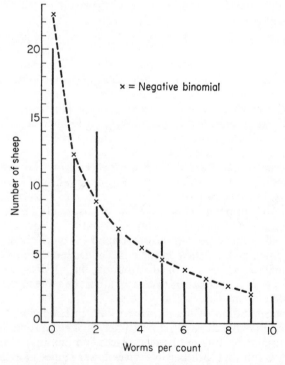

FIG. 107. Frequency distribution of eggs of a parasitic nematode in standard samples from faeces of sheep in Scotland. Values calculated from a negative binomial are shown for comparison.

probably subjected to a more uniform infestation, resulting in a more random distribution of worms."

MALARIA PARASITES IN THE BLOOD OF INDIAN CHILDREN

As the level of parasitism in most of the examples already discussed has been low, it is important to study a case where the infestation is high, to see if and how the pattern changes.

In 1924, Christophers gave details of the number of malarial parasites found in standard volumes of blood (1 cc) from thirty-one children under 5 years old, all infected with malaria, from India. The figures in order of increasing abundance are shown in Table 120 A. They range from 100 to over 100,000. In the lower part, B, of the table the cases are grouped into × 3 log classes (see p. 9) and the inset in Fig. 108 shows that the frequency distribution approximates to a symmetrical normal distribution with this log scale of units per group. The table also shows the accumulative total of cases, and this expressed as percentages of the whole, from which Fig. 108B has been drawn on probability paper. It will be seen how closely the values of the accumulated totals fit to a straight line indicating a log-normal distribution with a median log of 3.5 and a standard deviation of 0.75. Converting this to an arithmetic scale indicates that the frequency distribution of the malaria parasites in these blood samples fits to a log-normal distribution with a median of approximately 3000 × /÷ 5.75. The range between one S.D. above and below the median (within which approximately 68 per cent of the cases

TABLE 120. *Numbers of malarial parasites per unit volume of blood from thirty-one Indian children*

A. Actual observed counts:

100	455	1920	6741	16,855
140	770	2280	7609	18,600
140	826	2340	8547	22,995
271	1400	3672	9560	29,800
400	1540	4914	10,516	83,200
435	1640	6160	14,960	134,232
		6560		

B. Classification on × 3 log classes:

Class	No. of cases	Accum. total	Accum. total as %
V	1	1	3.2
VI	3	4	12.9
VII	5	9	29.0
VIII	6	15	48.4
IX	8	23	74.2
X	5	28	90.3
XI	2	30	96.8
XII	1	31	100

should occur) is thus 520–(3000)–17,200 parasites per standard volume of blood. The arithmetic mean number of parasites per sample is 12,500, with only 7 values above and 24 below.

The question of diversity does not arise, as the sampling was made by groups.

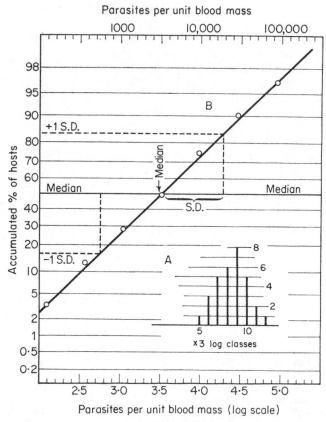

FIG. 108. Frequency distribution of Indian children with different levels of infection with malarial parasites per unit volume of blood. A shows the distribution with × 3 log classes of infection; B shows a "probability" relation indicating a very close fit to a log-normal distribution.

FREQUENCY DISTRIBUTION OF HUMAN DISEASES

As a contrast to the distribution of parasites of one particular kind on a number of individual hosts, it is interesting to consider some data presented by Herdan (1957) on the frequency distribution of a number of human diseases with different numbers of diagnoses in the hospitals in England and Wales for the year 1949. Herdan analysed his figures separately according to the sex of the patient, and according to whether the patients were in teaching or in regional board hospitals. He was chiefly concerned with a measurement of diversity as a characteristic of the distribution, so as to compare the uniformity or diversity of the diagnoses under different medical conditions. He suggests that the distribution appears to be between the log normal and the log series.

As an example of his results, Table 121 and Fig. 109 show the information available about diseases of male patients in teaching hospitals, in which, with 94,936 patients, diagnoses were made of 718 diseases. Two diseases were diagnosed over 4000 times each, and 120 diseases were diagnosed five times or less.

TABLE 121. *Frequency distribution of human diseases in hospitals in Britain with different numbers of diagnoses*

Frequency of diagnosis	No.	No. of diseases Accumulated total	As % of total diseases
1–5	120	120	16.7
6–10	55	175	24.3
11–25	103	278	38.7
26–50	114	392	54.6
51–100	115	507	70.6
101–250	117	624	86.9
251–500	56	680	94.7
501–1000	28	708	98.6
1001–2000	6	714	99.4
2001–4000	2	716	99.7
over 4000	2	718	100

If the 94,936 diagnoses were distributed among 719 diseases in the form of a log series, it can be shown that the value of n_1 should be approximately 106, and the sum of $n_1 - n_5$ should be about 240. Thus the observed number of 120 is only half of the expected (see the two points on Fig. 109).

To test the log-normal distribution, the accumulated percentage of the total diseases on a probability scale (see Table 121) has been plotted in Fig. 109 against the frequency of diagnoses per disease on a log scale. The resulting relation is very close to a straight line except for the rarer diseases which give slightly too high a proportion, supporting Herdan's suggestion that the distribution is between the log normal and the log series; but it is much closer to the former. The straight line drawn as a close fit to the points indicates a log normal with the median at 1.55 on the log scale, which equals 35.5 diagnoses per disease. The standard deviation is ± 0.65 on the log scale or $\times / \div 4.7$ on the arithmetic scale.

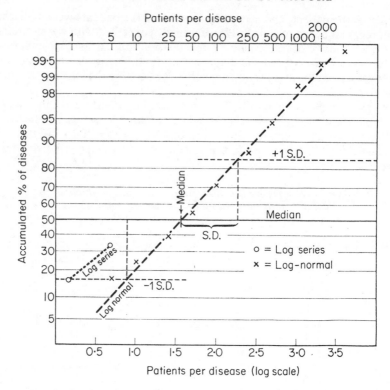

Fig. 109. Frequency distribution of human diseases with different numbers of diagnoses in hospitals in Britain.

DISTRIBUTION OF LARVAE OF WHEAT MIDGES (CECIDOMYIDAE) IN GRAINS OF WHEAT IN ENGLAND

In 1932, Barnes published figures showing the frequency distribution of the larvae of two Wheat Blossom Midges (Diptera Cecidomyidae) in flowers and grains of wheat at Rothamsted Experimental Station, about 25 miles north of London. Two species were studied, the first of which, *Sitodiplosis mosellanae* Gehin lays its eggs nearly always singly; the second *Contarinia tritici* (Kirby) tends to lay a number at a time. The eggs are laid in the flowers, and the larvae feed on the developing grains.

Over a period of 5 years about 125,000 grains were examined and in the case of *S. mosellanae* 14,978 larvae were found distributed in 10,605 grains, with an arithmetic mean of 1.41 larvae per infested grain. For *C. tritici* the corresponding figures were 68,839 larvae in 5560 grains, with an arithmetic mean of 12.38 larvae per infested grain. Thus more grains were infested with *S. mosellanae* than with *C. tritici*, but there were many more larvae of the latter species.

The frequency distribution of infestation for each species is shown in Tables 122 and 123 and in Fig. 110.

The question of the zero term, the number of grains without larvae, is difficult to decide. If all the grains were considered part of the same sample the zero terms would be approximately 114,400 for *S. mosellanae* and 119,500 for *C. tritici*, but it has to be remembered that many of the flowers, either by age or by situation, might not have been suitable or available for oviposition.

The distribution of *C. tritici*, with its peak on the arithmetic scale at n_{11}, is undoubtedly some form of contagious distribution, owing to the eggs being

TABLE 122. *Distribution of larvae of the blossom midge*, Contarinia tritici (*Cecidomyidae*), *in grains of wheat in England*

Larvae per grain	No. of grains	Larvae per grain	No. of grains	Larvae per grain	No. of grains	Larvae per grain	No. of grains	Larvae per grain	No. of grains
1	212	15	215	29	40	43	4	57	3
2	246	16	228	30	30	44	5	58	2
3	243	17	166	31	25	45	5	64	1
4	246	18	165	32	24	46	4	65	1
5	277	19	129	33	26	47	7	66	1
6	270	20	98	34	12	48	5	68	1
7	281	21	96	35	13	49	2	69	1
8	272	22	74	36	14	50	2	75	1
9	305	23	64	37	12	51	1	81	1
10	305	24	55	38	8	52	4	83	1
11	322	25	62	39	11	53	1	85	2
12	267	26	54	40	4	54	1	88	1
13	269	27	34	41	9	55	1	91	1
14	280	28	30	42	6	56	2		

Total = 68,835 larvae in 5560 grains

Log. × 3 classes		Square-root classes			
1	212	1	106.0	6	326.0
2	735	2	718.0	7	88.0
3	2568	3	1375.5	8	15.5
4	1969	4	1924.5	9	6.0
5	76	5	992.0	10	5.5

TABLE 123. *Distribution of larvae of the blossom midge* Sitodiplosis mosellana (*Cecidomyidae*) *in grains of wheat in England*

Larvae per grain	No. of grains	Larvae per grain	No. of grains	Larvae per grain	No. of grains	Larvae per grain	No. of grains	Larvae per grain	No. of grains
1	7655	3	634	5	59	7	13	9	1
2	2011	4	199	6	29	8	3	10	1

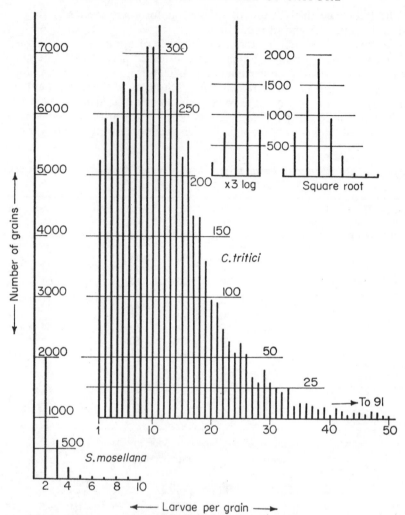

Fig. 110. Frequency distribution of seeds of wheat with different numbers of larvae of two wheat-blossom midges (Diptera): *S. mosellana* which lays its eggs singly, and *C. tritici*, which lays its eggs in groups. The data for *C. tritici* are also given with square-root and log transformation of the number of larvae per grain.

laid in groups. Log and square-root transformations are shown in both table and figure, but neither give a symmetrical distribution.

For *S. mosellana* a Poisson series, with the same number of grains and larvae, gives 111,600 grains with no larvae (as compared with the observed number 114,400), 12,500 with one larva (observed 7655), 750 with two larvae (observed 2011) and three with three larvae (observed 634). Thus the calculated values include too many free grains and too few with several larvae. There is a departure from random in the direction of slight aggregation.

These figures require further analysis before any particular pattern can be suggested.

DISTRIBUTION OF SCALE INSECTS ON LEAVES OF CITRUS IN NEW ZEALAND

Spiller (1948) discussed the distribution of the scale insect *Aonidiella aurantii* (Coccidae) on leaves of citrus trees in New Zealand and, taking a single leaf as the sampling unit, obtained the data shown in Table 124 and Fig. 111. It will be seen that a random sample of 720 leaves had on them 7520 scales, with an arithmetic mean of 10.45 scales per leaf, and included eighty-nine leaves free from scales.

Spiller pointed out, by means of a probability diagram, that the results could be closely represented by a log-normal distribution, truncate at the lower end, with the missing portion included in the zero term.

I have summarized his data in × 3 log classes, and the diagram shown in

TABLE 124. *Distribution of scale insects* (Aonidiella aurantii) *on leaves of citrus in New Zealand*

Scales per leaf	No. of leaves	Scales per leaf	No. of leaves	Scales per leaf	No. of leaves
0	89 (89)	22	7	47	1
1	79 (79)	23	4	50	1
2	78	24	7	52	2
3	58	25	5	53	1
4	51 (187)	26	3	55	1
5	30	27	3	57	2
6	39	28	8	59	1
7	34	29	2	60	1
8	27	30	5	62	2
9	18	31	2	65	1
10	13	32	2	74	1
11	12	33	2	78	1
12	19	34	2	80	1
13	16 (258)	35	1	81	1
14	13	36	1	88	1
15	6	37	4	94	1
16	12	38	1	96	2
17	6	39	2 (122)	103	1
18	5	41	2	105	1 (33)
19	8	42	2	185	1
20	6	44	5	241	1 (2)
21	5	45	1		

Total leaves		720
Total scales		7520
Arithmetic mean scales per leaf		10.45

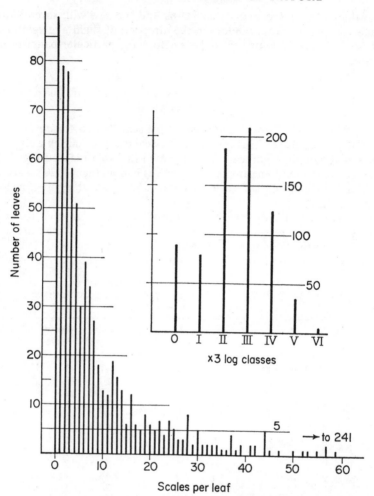

FIG. 111. Frequency distribution of scale insects (Coccidae) on leaves of citrus in New Zealand (Spiller).

Fig. 112 indicates a good fit to a log normal with a median at 0.67 (4.6 scales per leaf) and a S.D. of 0.55.

Spiller also pointed out that by increasing the size of the sampling unit, e.g. by taking two or more leaves together, it is possible to eliminate the zero term, and to get an untruncated log normal, except in so far as theoretically there is no upper or lower limit to any normal distribution. For example, using samples of twenty leaves together, he was able to show that the populations on different trees had a common variance, and thus normal statistical methods could be used.

Thompson (1950) returned to this problem and showed that by the use of a method of equating the moments of the sample distribution to those of the theoretical model, he gets, for Spiller's data, a median of 0.7 and a S.D. of 0.55. This is a more accurate method than the graphical one, but the difference is surprisingly small.

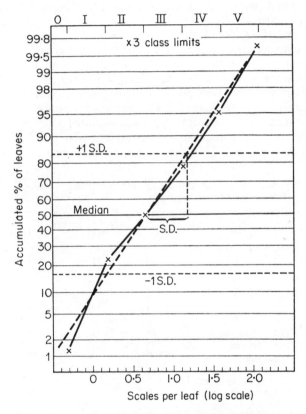

FIG. 112. Scale insects on citrus leaves in New Zealand. Data as in Fig. 111 on "probability diagram" showing a moderately close fit to a log-normal distribution.

THE DISTRIBUTION OF LARVAE OF WHITE FLIES (ALEYRODIDAE) ON OAK LEAVES IN SCOTLAND

The Aleyrodidae, or White Flies, are a group of Homopterous insects related to the Aphidae and Coccidae. In the white flies the adults are winged in both sexes, but all the immature stages, except the just-hatched larva, are sedentary and inactive. Once the young larva has settled down to feed, no further change of position occurs until the adult emerges. The extent to which the young larva can crawl between hatching and settling has not been studied, but it seems unlikely that it will move from the leaf on which the egg was laid.

Pealius quercus is widely distributed in Britain and has a single brood each year. Adults are found on the leaves of oak (Quercus spp) in the late spring, and the fully fed larvae fall to the ground in the autumn still attached to the leaf. The adults emerge from these in the following spring.

At the beginning of June 1961 I found adults common on oak leaves in mixed woodland round Loch Earn, in Perthshire, Scotland; and in October of the same year the fully grown larvae were common in the same district. Between 9 and 12 October I made two collections of leaves near St Fillans. The oak leaves were just beginning to fall. The first collection was a single branch about 5 ft from the ground which was found to bear seventy-three leaves. The second, made a few days later, consisted of four or five small branches from the same tree, but at a slightly higher level, 6 to 8 ft: these bore altogether 262 leaves. Each leaf was examined separately and the number of white-fly larvae, ranging from 0 to 5, was recorded as shown in Table 125. Owing to the very close similarity between the two samples it was considered permissible to combine the two into a single large sample (Fig. 113).

TABLE 125. *Distribution of larvae of the white fly* (Pealius quercus) (*Aleyrodidae*) *on oak leaves in Scotland*

No. per leaf	No. of leaves 1st coll.	2nd coll.	Total
0	57	210	267
1	13	38	51
2	2	7	9
3	0	5	5
4	1	1	2
5	—	1	1
Total with insects	16	52	68
Total leaves	73	262	335
Total insects	21	76	97
% leaves infected	21.9	19.9	20.3
Average insects per leaf			
All leaves	0.29	0.29	0.29
Infected leaves only	1.3	1.5	1.43

It will be seen that with ninety-seven insects spread over 335 leaves, 267 or just over 80 per cent of the leaves are free from infection. If the ninety-seven insects had been scattered at random over the leaves the resulting frequency distribution of leaves with 0, 1, 2, etc., insects would follow a binomial distribution of which the first few terms are shown in Table 126. In this, the number of leaves without insects would be 251 or 75 per cent, while only 0.8 of a leaf would have more than two insects as compared with the observed 8. Thus the observed results show more leaves without insects and more with several than expected on a random distribution. There is definite evidence of aggregation which must have some biological explanation. The results are shown diagrammatically in Fig. 113.

One possible factor leading to this would be that some of the oak leaves may have developed after the egg-laying period was over. A second explanation would be inactivity in the egg-laying females. It is known that in some species of white flies the females seldom fly except when disturbed, and in a few extreme cases the female lays a number of eggs without removing its

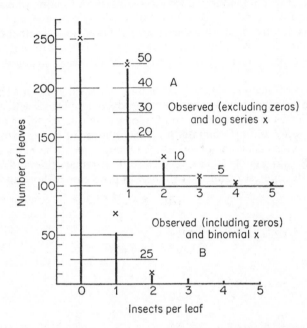

Fig. 113. Frequency distribution of last-stage larvae of a white fly (Aleyrodidae) on leaves of oak in Scotland, showing for comparison numbers estimated from a binomial (random) distribution and from the logarithmic series.

TABLE 126. *Comparison of observed distribution of white flies on oak leaves with estimates from random (binomial) and log-series distributions*

Larvae per leaf	Observed	Calculated random binomial	Calculated log series excluding zeros
0	267	250.6	—
1	51	73.1	49.47
2	9	10.5	12.12
3	5⎫	⎫	3.96⎫
4	2⎬8	⎬0.8	1.46⎬6.48
5	1		0.57
over 5	0⎭	⎭	0.42⎭

sucking mouth parts from the leaf. In these species the eggs, separated only by the rotation of the body axis, are laid in an arc of a tiny circle. Such a habit would lead to some form of contagious distribution. Unfortunately very little is at present known of the habits of *P. quercus* in this direction.

After the eggs hatch, any wandering of the young larvae which took them to another leaf would tend to reduce the aggregation. Finally, when the larvae have settled down they will be subject to attack by parasites and predators, and any selection by these in favour of several hosts close together on a single leaf would also tend to reduce the aggregation. It is most important for this problem that observations should be made at intervals throughout the life cycle.

It is also possible to study the distribution by considering only the infected leaves. Here we have ninety-seven insects on sixty-eight leaves with an average of 1.43 per leaf. On the assumption of a logarithmic series this would give a value of $x = .49$ and an estimate of n_1, the number of leaves with only one insect, as 49.5 instead of the observed 51. Other terms of the series are shown in Table 126 and Fig. 113. The very close fit is remarkable and is another example of the interest of this distribution in problems of host and parasite.

Chapter 9

THE PROBLEM OF INTRA-GENERIC
COMPETITION

IT has generally been held by ecologists that the more closely species are related in their habits and ecological relationships the more they will compete with one another in the struggle for survival, particularly in small communities.

If this is correct, and if the classification of animals and plants bears any relation to their ecological reactions, then there should be greater competition between species more closely related systematically, and particularly between species of the same genus. It has, in fact, become almost an axiom of ecology that in small communities this results in the early elimination of all but one or two of any group of congeneric species in a developing community or association.

The problem is capable of statistical study, and two approaches are given below. The first assumes that the frequency distribution of genera with different numbers of species is closely represented by the logarithmic series, and makes use of the Index of Generic Diversity calculated from this (Williams, 1947 *b*). The second uses Simpson's Index which is independent of any assumption about the mathematical form of the distribution. Both lead to the same conclusion, which requires a considerable revision of our ideas on this problem.

The species in any group of animals or plants that are found living side by side in a relatively small community have been selected in the course of time from all the species in the surrounding areas which have been able to reach the smaller area in question. Those that survive are those that are capable of existing in the physical environment of the area, and also in association with, or in competition with, the other members of the community.

Such a natural selection could conceivably be brought about in three different ways: (1) without reference to the generic relations of the species, (2) more or less against further species in the same genus, and (3) more or less in favour of further species in the same genus. The first would be a truly random selection of species, so far as generic relation is concerned. Extremes of either (2) or (3) would result in only one species in each genus on the one hand, or in all the available species in each genus, on the other hand, being represented. We know, however, that biologically neither of these extremes is correct.

On many occasions it has been assumed that if the average number of species per genus in a smaller community is less than that in the larger surrounding area, this alone is evidence of selection against congeneric species. This is, however, a statistical error as, with a truly random sample of a small number of species from a larger population, there is always a reduction in the average number of species per genus.

It is therefore important to consider in detail what happens when a selection of a relatively small number of species is made from a larger fauna or flora, without reference to their generic relations, as a true interpretation can only be obtained by comparing the observed results with those obtained from a random sample. If the observed average number of species per genus is smaller than that found in a random sample, or the diversity is larger, then there is evidence of a selection against generically related species; if the observed average is larger than would be expected by random sampling, or the diversity smaller, then the selection has been in favour of species in the same genus.

As several of the examples which will be discussed later relate to the British flowering plants, the classification of these can be taken to illustrate the results of random sampling, and also for comparison with the observed results in the field.

Table 127 shows some results of sampling by random selection of species from the British flowering plants as classified by Bentham and Hooker in 1912. The particular classification is of little importance provided that the same one is used throughout. They recognized 1251 species which they classified into 479 genera with an average of 2.61 species per genus. This, on the assumption of a logarithmic series, gives a generic diversity of 284. All ran-

TABLE 127. *Number of genera, expected from calculation and found by artificial sampling, in random samples of different numbers of species from the British flowering plants as classified by Bentham and Hooker to include 1257 species and 497 genera*

No. of sp. in random sample	Calculated from logarithmic series		Average of three artificial samples	
	No. of genera	Ratio sp./gen.	No. of genera	Ratio sp./gen.
1000	428	2.34	—	—
500	288	1.74	—	—
200	151	1.32	—	—
161	127.5	1.26	123.7	1.30
100	86	1.16	—	—
73	65	1.12	64.3	1.14
65	58.6	1.11	59.7	1.09
59	53.6	1.10	53.3	1.11
50	46	1.09	—	—
30	28.5	1.05	28.7	1.05
25	24.0	1.04	—	—
23	22.1	1.04	22.3	1.03
20	19.3	1.04	19.7	1.02
10	9.8	1.02	—	—

dom samples from such a population should have the same diversity and so it is possible to calculate the expected number of genera for a sample of any particular number of species; and this has been done for a number of samples ranging from ten to 1000 species as shown in the same table.

It will be seen that, in spite of the samples being random, the ratio of species to genera falls steadily from 2.61 in the total flora to 1.02 when only ten species are selected. It is easy to see that, in the limit, if only one species was selected it could belong to only one genus and the ratio would fall to the limit of 1.00. It should also be clear that the only way in which the ratio of species to genera could remain the same in a sample as in the original population would be if there was a random selection of complete genera and not of species.

It will be seen from the table that if (from this population with a diversity of 284) a sample of twenty-five species is selected the expected number of genera is twenty-three, and so only one pair of congeneric species would be expected to occur.

To check the above results, and particularly to get some estimates without assuming the existence of a log series, some artificial selections were made by putting 1251 numbered discs in a box and, after prolonged shaking, withdrawing samples of increasing size and recording the number of genera represented at different levels. The average results of three such selections are shown in Table 127 and it will be seen how very closely they correspond to the calculated values, particularly with thirty species selected where the artificial samples indicated 28.7 genera and the calculated 28.5, and with fifty-nine species, 53.3 in the samples and 53.6 calculated from the log series.

It is clear, therefore, that we have a reliable method of estimating

TABLE 128. *Observed number of species of flowering plants found in smaller and larger areas in Britain together with the number of genera expected from a random selection of the same number of species from the British flora*

	No. of sp.	No. of genera Observed	No. of genera Expected in random sample	Ratio sp./gen. Observed	Ratio sp./gen. Expected in random sample	Generic diversity observed
1 British flowering plants	1251	479	—	2.61	—	284
2 Scolthead Is., Norfolk	161	114	127	1.41	1.28	182
Rothamsted Expt., Stat., Herts. "Broadbalk Wilderness"						
3 Four surveys total	73	59	65	1.24	1.12	147
4 1913 only	65	52	59	1.25	1.11	134
"Park Grass"						
5 All plots	59	53	56	1.11	1.05	263
6 Plot 3, 1919	30	27	28.5	1.11	1.05	134
7 Plot 13, no lime	20	18	19.3	1.11	1.04	89
8 Plot 13, with lime	23	21	22.2	1.10	1.04	111

the number of genera to be expected when a random sample of a particular number of species is taken from a larger population in which the total number of species and genera is known.

To study the problems of intrageneric competition quantitatively in the field we require a number of cases where we know the number of species and the number of genera of any particular group of animals or plants in a larger area and also in one or more smaller areas within the larger. For example, if we knew the generic diversity of the flowering plants of Europe, of Britain, of a British county, and of a smaller area within that county (all using the same basic classification), we would have a series from which it should be possible to show any trend in diversity with decreasing size which might be significant.

It has not been possible to get such a complete series, but several examples of shorter series are given in Table 128 from the British flora and in Table 129 from British insects. For a fuller discussion see Williams (1947).

TABLE 129. *Observed numbers of species and genera of insects found in smaller and larger areas in Britain, together with the number of species expected from random samples of the same number of species from the total British fauna*

	No. of sp.	No. of genera	Ratio sp./gen.	Generic diversity smaller area	larger area
Macro-Lepidoptera					
1 Britain, Meyrick, 1895	788	212	3.72	—	96.5
2 Britain, Meyrick, 1927	806	218	3.70	—	96.6
3 Cambridge, Wicken Fen	368	135	2.73	76.4	—
4 Hertfordshire	561	186	3.02	99	—
Noctuidae (Caradrinidae) only					
5 Britain, Meyrick, 1895	273	39	7.00	—	12.7
6 Britain, Meyrick, 1927	290	45	6.44	—	14.9
7 Cambridge, Wicken Fen	146	29	5.04	10.8	—
8 Hertfordshire	180	33	5.15	13.2	—
Plusiidae only					
9 Britain, Meyrick, 1895	54	22	2.45	—	14.0
10 Cambridge, Wicken Fen	24	12	2.00	9.3	—
Coleoptera					
11 Britain, 1904	3268	804	4.60	—	341
12 Berkshire, Windsor Forest	1825	553	3.30	275	—
Heteroptera, Miridae only					
13 Britain, 1943	186	76	2.21	—	48
14 Hertfordshire, 1945	127	60	2.12	43.7	—

I. BRITISH FLOWERING PLANTS

(A) SCOLTHEAD ISLAND, NORFOLK

Chapman (1934) gave a list of the flowering plants of Scolthead Island, which is an area of about 1.5 sq. miles of sand-dunes just off the north coast of Norfolk. His list, when altered to the nomenclature of Bentham and Hooker as stated above, shows 161 species in 114 genera, with an average of 1.41 species per genus. The generic diversity is 182 as compared with 284 for the whole British flora. In a purely random sample of 161 species from the flora (see Table 128, *2*) we would have expected to find 127 genera represented. Thus the observed number of genera is smaller than the calculated, indicating a selection in favour of generically related species rather than against them.

(B) BROADBALK WILDERNESS, ROTHAMSTED EXPERIMENTAL STATION (TABLE 128, *3* and *4*)

Broadbalk Wilderness is a small area of about half an acre in the Experimental Station at Harpenden, Herts., which was allowed to go wild in the year 1882. Brenchley and Adam (1915) published a list of the flowering plants found in four surveys carried out at intervals between 1867 and 1913, which included a total of seventy-three species classified into fifty-nine genera. The average number of species per genus was thus 1.24 and the generic diversity 147.

If only the species observed in the 1913 survey are considered, the numbers are sixty-five species in fifty-two genera, with an average of 1.25 and a diversity of 134.

When these are compared with the total British flora, the average number of species per genus is smaller, as would be expected from the process of sampling, but the diversities are also both lower, indicating selection in favour of congeneric species. In fact, the number of genera expected in a random sample of seventy-three species from the British flora (see Table 128, *3*) is sixty-five, instead of the fifty-nine observed, and the average number actually found in three artificial random samples was 64.5, almost identical with the theoretical value.

(C) PARK GRASS PLOTS, ROTHAMSTED EXPERIMENTAL STATION

At this station there are a series of plots of grass, ranging from $\frac{1}{2}$ to $\frac{1}{8}$ acre in extent, each of which has been manured in a special way since 1856, but in which no further interference has been made in the natural flora which develops under such conditions of soil, climate and manuring, except to mow the grass once or twice a year.

Brenchley (1924) has given a number of details about the flora from which the information in Table 128, *5–8*, has been extracted. It will be seen that the total recorded flora of all plots in all surveys was fifty-nine species belonging to fifty-three genera, with an average per genus of 1.11 and a diversity of 263. Taking single plots in a single year Plot, 3 unmanured, gave identical numbers of species (30) and genera (27) in each of its two halves, limed and unlimed, with a generic diversity of 134. Plot 13, with farmyard manure and a much more luxuriant growth, gave twenty species in eight genera in the half without lime, and twenty-three species and twenty-one genera with lime. The generic diversities of both of these, at 89 and 111, are low, but it must be noted that with only twenty species the error of estimation is high.

All these plots give generic diversities below that of the total flora, indicating a selection in favour of congeneric species.

11. British Insects (Table 129)

(A) Macro-Lepidoptera

Meyrick (1895, 1927) published two books dealing with the classification of the British Lepidoptera, the edition of 1927 being used as the standard in the discussions below. In this he recognized 806 species of the so-called Macro-Lepidoptera in 218 genera, with an average of 3.7 species per genus, and a generic diversity of 99.6. In his earlier volume (1895) he recognized 788 species in 212 genera with an average of 3.72 species per genus and a diversity of 96.5. The differences between the two are almost negligible, and it is of interest to see how little his ideas on the scope of a genus had changed in the interval of 32 years.

Foster, in 1937, gave a list of the Macro-Lepidoptera of the county of Hertfordshire (836 sq. miles) in which he enumerates 561 species classified into 186 genera, giving 3.02 species per genus and a diversity of 99.

Farren, in 1936, gives a list of the Macro-Lepidoptera of Wicken Fen, an area of almost original fen-land covering about 730 acres in Cambridgeshire. He used Meyrick's 1895 classification and recognizes 368 species in 135 genera, with an average of 2.73 species per genus and a diversity of 74.6.

. It will be seen that in both the smaller areas the number of species per genus is below that of the British Isles, but also the generic diversities are lower, indicating fewer genera than would be expected by random sampling.

(B) Lepidoptera of the Family Noctuidae

Meyrick in 1927 recognized in the family Noctuidae (Agrotidae or Caradrinidae) 290 species in forty-five genera, an average of 6.44 species per genus and a diversity of 14.9 (Table 129, 6). Foster (1937), using the same classification, recognized for the county of Hertfordshire 180 species in thirty-three genera; 5.15 species per genus and a diversity of 13.2 (Table 129, 8).

Farren, in 1936, recognized for Wicken Fen 146 species in twenty-nine genera, which is five species per genus and a diversity of 10.8, but the classification he used was that of Meyrick's first volume (1895), which differed only slightly from the later edition in having 273 species in thirty-nine genera, with 7.00 species per genus and a diversity of 12.7 (Table 129, 7 and 5).

Once again there is a descending level of generic diversity as the area under consideration becomes smaller. In this case, however, the difference in diversity between the total British and the Hertfordshire lists is very small.

(C) Lepidoptera, Family Plusiidae (Table 129, 9 and 10)

In the family Plusiidae, Farren (1936) recognized twenty-four species in twelve genera, with an average of 2 species per genus and a generic diversity of 9.3. Meyrick, in his 1895 edition, as used by Farren, considered as British fifty-four species in twenty-two genera, giving an average of 2.45 species per genus, and a diversity of 14. Once again the results point in the same direction.

(D) BRITISH COLEOPTERA (TABLE 129, *11* and *12*)

Beare and Donisthorpe, in 1904, recognized as British 3268 species of beetles which they classified into 804 genera. This gives an average of 4.06 species per genus, and a generic diversity of 341 (Williams, 1944). Donisthorpe (1939) in a list of the Coleoptera of Windsor Forest, of about 6250 acres in Berkshire, recognized 1825 species in 553 genera with an average of 3.3 species per genus and a diversity of 275. Thus the beetles of the forest have a lower diversity than that of the larger area, indicating selection in favour of congeneric species.

(E) BRITISH HETEROPTERA, FAMILY MIRIDAE OR CAPSIDAE (TABLE 129, *13* and *14*)

China (1943) in a list of the Heteroptera of the British Isles records, in the family Miridae, 186 species in seventy-six genera, giving 2145 species per genus and a generic diversity of 48. In 1945, Bedwell, for the county of Hertfordshire, recognized 127 species in sixty genera with an average of 2.12 species per genus and a diversity of 43.7. This diversity is less than that for the whole fauna, but probably not significantly so. In any case there is no support for any change that would imply selection against congeneric species.

When the above information was first published in 1947 some criticisms were made, one of which was that the sizes of the smaller areas considered were still sufficiently large to have a complex environment and to contain a large number of "niches" (Bagenal, 1951).

The difficulty of obtaining any conclusive statistical evidence from small samples has already been mentioned. When a population is sampled, then the minimum size of sample that can be expected to contain a single pair of units from the same group must contain N units such that $N (N—1)/2$ equals the diversity of the population. For example, in the British flora, with a diversity of 284, a random sample must contain at least twenty-four species ($24 \times 23 \div 2$ possible pairs) before one pair of congeneric species would be expected to appear. The absence of congeneric species in a smaller sample would be no evidence of dispersal.

With a population with a diversity of 50 the minimum sample size would have to be ten species before any two could be expected to be congeneric.

The second main criticism was my assumption of the logarithmic series, although the frequency distribution of the genera with different numbers of species closely supported it, and although artificial random samples gave results so close to those calculated.

An opportunity occurred later (Williams, 1951c) to study statistically the distribution of a number of birds over a series of small ecological habitats in East Africa, in which the numbers of species per habitat ranged from one up to a maximum of twenty. It seemed possible that by adding together all the results we could obtain significant evidence even though the expected number of congeneric species might be small.

The opportunity was also taken to use Simpson's measure of diversity: the number of pairs of species which have to be selected in order to include on an average one pair from the same group. When this is calculated directly from the actual numbers of genera with different numbers of species in each habitat it is independent of any theory of the mathematical form of the frequency distribution.

In 1948, Moreau published a detailed account of the distribution of 172 species of birds, belonging to nine families of passerines, in the Usambara hills of north-east Tanganyika, with special reference to the occurrence of two or more species of the same genus co-existing in a single ecological assocation or community. He showed that there were a number of such cases, but that in many of them the related species have different habits which prevent them competing with each other; they occupy different niches within the community.

While admitting the probable correctness of these explanations, I thought that it would be of value to study his data from a statistical point of view, because no attempt had been made to see whether the number of such cases was larger or smaller than might be expected by chance; and also because Moreau's data, being the results of observations by a field ecologist who was familiar with his terrain and his birds, were likely to be as good as it is possible to get anywhere for such an analysis.

Moreau informed me that these nine families were selected for particular study because they were the largest families which he considered that he knew adequately. Within these families all species and genera were listed so that there was no selection for generic size. He recognized 172 species in ninety-two genera, and the following are the numbers of genera with different numbers of species in the whole population.

Species per genus	1	2	3	4	5	6	7	8	9	10
Number of genera	51	24	10	2	0	3	1	0	0	1

The names of the nine family or sub-family groups recognized by Moreau, together with the number of genera and species in each, are shown in Table 130.

TABLE 130. *Classification groups of East African birds discussed by Moreau*

	Species	Genera
1 Capitonidae	11	6
2 Pycnonotidae	15	6
3 Muscicapidae	18	12
4 Turdidae	20	12
5 Sylvidae	28	15
6 Prionopidae and Laniidae	21	14
7 Nectariniidae	17	6
8 Plocidae, Plocainae	21	7
9 Plocidae, Estrelidinae	21	14
Total	172	92

Moreau subdivided his area, which covered a total area of about 3000 sq. miles (two English counties), into thirty-two different ecological types as defined in Table 131; and Table 132 shows the distribution of the species, genera and families represented in each habitat. One or two misprints in the original tables have been corrected with the co-operation of Mr Moreau.

A preliminary study of the total population of 172 species in ninety-two genera and nine classification divisions gives us the following information:

(1) The total number of ways in which two species can be selected from 172 is $(172 \times 171)/2 = 14,706$.

(2) The number of ways in which two species can be selected so as to belong to the same genus is:

$$(24 \times 1) + (10 \times 3) + (2 \times 6) + (3 \times 15) + (1 \times 21) + (1 \times 45) = 177.$$

TABLE 131. *Classification by Moreau of the habitats of birds in East Africa, used for the discussion on intra-generic competition*

---

1. Lowland (0–2500 ft)

(a) Rain forest	(1) Tree-tops
	(2) Mid-stratum
	(3) Ground-stratum
	(4) Edges
(b) Riverine forest	(1) Trees
	(2) Ground-stratum
(c) Wooded grassland	(1) Grass
	(2) Trees (deciduous)
	(3) Low semi-evergreen bush
	(4) Tall clumps, semi-green bush
(d) Semi-desert thorn country	(1) Trees and bushes
	(2) Ground, including woody herbage
	(3) Riverine strips
(e) Induced vegetation	(1) Trees and tall bushes
	(2) Dense low bush
	(3) Herbaceous cover
	(4) Scanty cover
(f) Swamp	

2. Intermediate level (2500–4500 ft)

(a) Rain forest	(1) Tree-tops
	(2) Mid-stratum
	(3) Ground-stratum
	(4) Edges
(b) Grassland	
(c) Induced vegetation	(1) East Usambara (humid)
	(2) West Usambara (semi-humid)
(d) Swamp	

3. Highland (4500–7500 ft)

(a) Rain forest	(1) Tree-tops
	(2) Mid-stratum
	(3) Ground-stratum
	(4) Edges
(b) Moorland	
(c) Induced vegetation	

(3) The number of ways in which a pair of species can be selected so as to belong to the same family group is 1647, which includes the 177 cases where they belong to the same genus.

Thus of the 14,706 ways in which two species can be selected from the population:

in 177 they will belong to the same genus;

in 1470 they will be in the same family but not in the same genus;

in 1647 they will be in the same family group, including those in the same genus;

TABLE 132. *The number of genera of birds with different numbers of species in the thirty-two ecological habitats recognized by Moreau in East Africa*

Habitat group	No. of species	No. of genera with 1 species	2 species	3 species	4 species	5 species	No. of genera	No. of families
1 a 1	9	7	1	—	—	—	8	5
2	7	7	—	—	—	—	7	5
3	7	5	1	—	—	—	6	4
4	13	11	1	—	—	—	12	6
b 1	1	1	—	—	—	—	1	1
2	4	1	0	1	—	—	2	2
c 1	7	3	0	0	1	—	4	3
2	13	11	1	—	—	—	12	6
3	7	5	1	—	—	—	6	4
4	6	6	—	—	—	—	6	4
d 1	14	14	—	—	—	—	14	6
2	14	12	1	—	—	—	13	5
3	15	5	1	1	0	1	8	4
e 1	11	8	0	1	—	—	9	5
2	9	7	1	—	—	—	8	5
3	13	6	2	1	—	—	9	3
4	3	3	—	—	—	—	3	2
f	4	4	—	—	—	—	4	2
2 a 1	6	4	1	—	—	—	5	4
2	9	9	—	—	—	—	9	5
3	13	10	0	1	—	—	11	6
4	12	10	1	—	—	—	11	6
b	5	5	—	—	—	—	5	2
c 1	20	18	1	—	—	—	19	8
2	12	12	—	—	—	—	12	8
d	3	3	—	—	—	—	3	2
3 a 1	5	5	—	—	—	—	5	4
2	6	4	1	—	—	—	5	4
3	10	6	2	—	—	—	8	5
4	7	7	—	—	—	—	7	5
b	5	5	—	—	—	—	5	36
c	10	10	—	—	—	—	10	
Av. per habitat	8.75	7.00	0.50	0.16	0.03	0.03	7.72	4.37

Av. no. of species per genus per habitat = 1.132.

in 13,059 they will *not* be in the same family, and in 14,529 they will not be congeneric.

Therefore a randomly selected pair will be:

in the same genus once out of 83.1 selections;
in the same family group once out of 8.9 selections.

These numbers 83.1 and 8.9 are Simpson's measures of Generic and Family diversity.

Similarly it can be shown that:

for three species the chances that they will be congeneric are 233 out of 833,340, or one out of 3577,

for four species the chances that they will be congeneric are 292 out of 35,208,615, or one out of 120,577,

and for five species the chances are one in just over 3 million.

For an analysis of the frequency of genera with more than one species in a single habitat let us take as an example Moreau's habitat "Induced Vegetation, Herbaceous Cover" (1, e, 3), which has thirteen species in nine genera of which three genera have two species and one has three. The thirteen species contain $\frac{1}{2}$ (13 × 12) = seventy-eight different pairs. Of these two pairs in the two genera with two species each, and three pairs in the genus with three species are congeneric. We have already seen, however, that a random selection from the whole population gives only one congeneric pair in eighty-three, so that this particular habitat has about five times as many congeneric pairs as would be expected by chance.

For the same habitat the number of possible groups of three species is (13 × 12 × 11) ÷ (2 × 3) = 286, and the actual number of congeneric groups of 3 is one only, from the single genus containing three species. The expected frequency is, however, only one in 3577.

As it is not desirable to argue from single cases, Table 133 has been prepared to show similar analyses for each of the thirty-two habitats for groups of two, three, four and five species, and giving at the bottom the total possible groups and the actual congeneric groups that are found in all the single habitats together. The final results for all the thirty-two habitats are as follows:

(1) Out of 1372 possible pairs of species within a habitat, forty-seven are congeneric. The number expected by random selection is 1372/83.1 = 16.5.

(2) Out of 4849 possible groups of three species selected within a habitat, nineteen are congeneric. The expected number by random selection is 4849/3775 = 1.35.

(3) Out of 13,427 possible groups of four species within a habitat, six are congeneric. The expected number by random selection is 13,427/120.577 = 0.11.

(4) Out of 30,713 possible groups of five species one is congeneric. The expected number is 30,713/3 million = 0.01.

There are thus, within the single habitat associations, three times as many congeneric pairs of species, fourteen times as many groups of three, fifty

TABLE 133. *The number of species of birds in each of Moreau's thirty-two habitats together with the number of different groups of two, threee, four and five species which it is possible to select within each habitat—and the number of these that would be congeneric*

Habitat	No. of species	Pair of species Possible	Con-generic	Three species Possible	Con-generic	Four species Possible	Con-generic	Five species Possible	Con-generic
1 a 1	9	36	1	84	0	126	0	126	0
2	7	21	0	35	0	35	0	21	0
3	7	21	1	35	0	35	0	21	0
4	13	78	1	286	0	715	0	1287	0
b 1	1	0	0	0	0	0	0	0	0
2	4	6	3	4	1	1	0	0	0
c 1	7	21	6	35	4	35	1	21	0
2	13	78	1	286	0	715	0	1287	0
3	7	21	1	35	0	35	0	21	0
4	6	15	0	20	0	15	0	6	0
d 1	14	91	0	364	0	1001	0	2002	0
2	14	91	1	364	0	1001	0	2002	0
3	15	105	14	455	11	1356	5	3003	1
e 1	11	55	3	165	1	330	0	462	0
2	9	36	1	84	0	126	0	126	0
3	13	78	5	286	1	715	0	1287	0
4	3	3	0	1	0	0	0	0	0
f 4	4	6	0	4	0	1	0	0	0
2 a 1	6	15	1	20	0	15	0	6	0
2	9	36	0	84	0	126	0	126	0
3	13	78	3	286	1	715	0	1287	0
4	12	66	1	220	0	495	0	792	0
b	5	10	0	10	0	5	0	1	0
c 1	20	190	1	1140	0	4845	0	15,504	0
2	12	66	0	220	0	495	0	792	0
d	3	3	0	1	0	0	0	0	0
3 a 1	5	10	0	10	0	5	0	1	0
2	6	15	1	20	0	15	0	6	0
3	10	45	2	120	0	210	0	252	0
4	7	21	0	35	0	35	0	21	0
b	5	10	0	10	0	5	0	1	0
c	10	45	0	120	0	210	0	252	0
Total		1372	47	4829	19	13,427	6	30,713	1

times as many groups of four species, and 100 times as many congeneric groups of five species, as would be expected in samples from the same populations taken without reference to generic relations.

The same form of argument can be applied also to the relative frequency of species belonging to the same family (Table 130). Out of the 14,706 ways of selecting two species at random there are 1647 pairs belonging to the same family. One would therefore expect one pair out of 8.9 to belong to the same family.

The total number of possible pairs within the single habitats is 817 and the number of these that are within the same family is 277: the expected number by random selection is 91.5. There are thus three times as many pairs of species in the same family within single habitats as would be expected if there was no selection in respect of family relationship. Therefore, in this particular set of data, there is evidence of a higher proportion of species both in the same genus and in the same family than would be expected by chance selection.

I submitted the above analysis to Mr Moreau for comment and he replied that habitat 1, d, 3, "lowland, semi-desert, riverine strips", was occupied only during that part of the year when food appeared to be superabundant. It is therefore not so clear cut as the others, and species may move in for feeding purposes only.

A high proportion of the congeneric species are found in this habitat, but I am not sure that it is desirable to remove it from consideration, as competition for food is one of the more definite types of interspecific competition. I have, however, recalculated all the above figures relating to congeneric species leaving out habitat 1, d, 3.

In this area seven species occur that are not mentioned elsewhere. This reduces the total number of species to 165 and also reduces the size of several genera. The following are the results:

I. For pairs of species

Whole fauna:
    Total possible pairs 13,530
    Congeneric pairs   153
    Chance of a random pair being congeneric is 1 in 88.4
Single habitats:
    Total possible pairs 1267
    Congeneric pairs   33
        or 1 congeneric pair in 38
    This is 2.3 times as many as expected by chance selection.

II. For groups of three species

Whole fauna:
    Total possible threes 735,130
    Congeneric threes   165
        or 1 in 4460
Single habitats:
    Possible threes   4374
    Congeneric threes   8
        or 1 in 547
    This is eight times the expected frequency.

III. For groups of four species

Whole fauna:
    Possible fours   29,772,765
    Congeneric fours   151
        or 1 in 197,171
Single habitats:
    Possible fours   12,071
    Congeneric fours   1
    This is fifteen times the expected frequency.

Thus the effect of eliminating habitat 1, d, 3 is to reduce the extent of the excess of observed over calculated groups of congeneric species, but the excess remains with no possible indication of a selection against congeneric species.

The present situation can therefore be summarized as follows:

(1) In order to continue to exist in any particular area an animal or plant species must be able to hold its own under two sets of environment conditions. First the physical factors, which include soil conditions, drainage, etc., as well as the variations and particularly the extremes of the weather. Secondly, the biological factors, which include the struggle with other species for food and to avoid being eaten.

(2) There is a widely accepted, but by no means proved, theory that species which are more closely related taxonomically, as, for example, those in the same genus, are more similar in their food requirement and other habits, and hence will compete more severely with each other than with more distantly related species. This would result in a greater likelihood of the reduction or elimination of one of a pair of species when the pair are from the same genus than when not so. Evidence has been sought for this in the pattern of the species–genus relation in small communities.

(3) This present statistical investigation fails to find any support for the idea that in small communities there are fewer congeneric species (and therefore more genera for the same number of species) than would be expected in a random sample of the species in the surrounding areas. In fact, the evidence points very definitely the opposite way, and shows that in nearly every case examined there are more congeneric species and fewer genera than would be expected by chance.

(4) If this is so, then either the conception that biological competition is greatest between congeneric species is incorrect, or there is some more dominating influence in the opposite direction overshadowing any effect of biological competition.

(5) It is clear from a wide study of any group of animals or plants that there are many genera which, though differentiated basically on morphological characters, have very definite relations to physical factors. There are genera which are typical of the tropics, of the arctic regions, of the desert, of rapidly running water, and so on. It follows from this that if in any particular environment one species of a particular genus can survive, then other species of the same genus are, on the average, more likely to be suited to that physical environment than if they belonged to a more distantly separated genus.

(6) Therefore it would appear that the actual conditions found in wild populations—an excess of congeneric species—may well be due to their adaption to similar physical conditions being a sufficient advantage to offset any drawback due to competition for similar food or living space.

(7) Thus the results do not exclude the possibility of greater biological competition between closely related species. They suggest, however, that too much emphasis has been put in the past on the importance of the biological environment in the determination of survival, and too little on the role of the physical environment.

Chapter 10

SOME RELATED PROBLEMS IN STATISTICAL ECOLOGY

GENERA, FAMILIES AND SPECIES OF BIRDS: THE TRIANGULAR RELATION

IT sometimes happens that the classification of units into groups with different numbers of units per group can be carried out at several different levels in the same set of data. For example, in the classification of animals or plants the individuals can be grouped into species, these species into genera, the genera into families, and so on as discussed in Chapter 6.

If we consider the birds of the British Isles there are fifty-nine families containing 191 genera, and these genera contain 426 species. It has already been shown (p. 133) that the distribution both of families with different numbers of genera, and of genera with different numbers of species follows very closely a logarithmic series. It is, therefore, of some interest as to what would be the distribution of families with different numbers of species, omitting the intermediate class of genera.

The distribution of families and genera and of genera and species will be found in Table 58, p. 134; that of families and species is shown in Table 134.

TABLE 134. *Distribution of families of British birds with different numbers of species*

Species per family	Non-Passerine	Passerine	Total	Species per family	Non-Passerine	Passerine	Total
1	11	5	16	12	1	—	1
2	6	4	10	16	1	—	1
3	3	1	4	22	1	—	1
4	7	1	8	25	—	1	1
5	2	2	4	27	1	—	1
6	—	—	—	33	—	1	1
7	1	—	1	38	—	1	1
8	1	2	3	41	1	—	1
9	—	1	1	42	1	—	1
10	2	1	3	—	—	—	—

	Non-Passerine	Passerine	Total
Total families	39	20	59
Total species	265	161	426
Species per family	6.79	8.05	7.22
"x"	.954	.964	.958
Estimated n_1	12.1	5.8	17.7
Diversity	12.7	6.0	18.5

Figure 114 shows these results diagrammatically, together with the two other distributions for comparison.

The frequency distribution is once again of the hollow-curve type and from the close relation between the observed n_1 and that calculated from the log series it would appear that this new distribution is also very close to the logarithmic series, similar to those found with the intermediate classes.

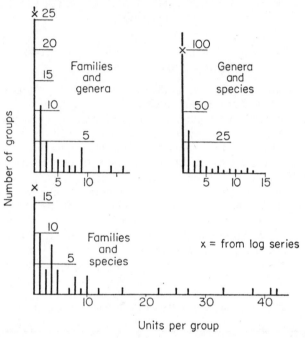

FIG. 114. Interrelation, in the classification of the British birds, of the numbers of species per genus and per family, and of the number of genera per family. Also shown is the first term (groups with one unit) for each relation calculated on a logarithmic series.

DISTRIBUTION OF GYRINIDAE (COLEOPTERA) IN SOME CANADIAN LAKES IN RELATION TO THEIR ABUNDANCE

In 1955, Robert discussed the distribution of fifteen species of water-beetles of the genus *Gyrinus* in thirteen lakes in the National Park of Mont Tremblant, which is in the province of Quebec, about 75 miles N.W. of Montreal. His original data were given in the form of percentages of each species of the total number of individuals found in each lake. As he also gave the total individuals, I have converted back the percentages to numbers of individuals. The results are shown in Table 135. The numbers for Lac François have been corrected by the author from a misprint in the original.

TABLE 135. Distribution of fifteen species of water-beetles (Gyrinidae) in thirteen lakes in Mont Tremblant National Park, Canada

Lake / Species	1 Lauzon	2 Des fammes	3 A Mousse	4 En coeur	5 Rocheux	6 Des sables	7 Mallard	8 Rubaniers	9 François	10 Boivin	11 St Louis	12 Des diable	13 Russeau boivin	Total individuals	Lakes ex. 13
1 Lugens	276	240	381	104	19	104	159	26	35	—	25	44	14	1427	12
2 Affinis	46	67	11	1	2	15	17	1	1	6	8	21	155	337	11
3 Latilimbus	2	42	2	3	—	—	60	3	217	28	—	39	10	392	11
4 Pugionis	—	13	3	—	25	8	19	2	—	—	—	—	7	72	6
5 Ventralis	—	6	—	—	—	—	—	—	—	—	—	—	—	6	1
6 Dichous	2	3	—	1	—	—	—	—	—	—	—	4	—	10	4
7 Fraternus	—	2	4	—	—	—	—	—	—	—	—	—	—	6	2
8 Pectoralis	2	3	134	3	242	—	4	—	—	—	—	—	—	388	6
9 Impressicollis	—	1	1	—	—	—	1	—	—	—	—	1	—	4	4
10 Frosti	1	1	—	—	—	—	—	—	—	—	—	—	—	2	2
11 Gehringi	—	—	47	—	—	—	—	—	—	—	—	—	—	47	1
12 Instabilis	—	2	4	4	—	—	—	2	4	—	—	—	30	72	5
13 Minutus	—	—	—	—	2	—	1	—	—	1	—	—	—	10	3
14 Bifarius	—	—	—	—	—	43	—	—	—	—	—	2	—	45	2
15 Lecontei	—	—	—	—	—	—	—	—	—	7	1	—	4	11	2
Total individuals*	329	380	587	116	290	170	261	34	257	42	34	111	220	2830	
Total species	6	11	9	6	5	4	7	5	4	4	3	6	6	15	
Diversity (Simpson)	1.4	2.3	2.1	1.2	1.4	2.3	2.3	1.3	1.4	2.1	1.3	3.2	1.9	3.25	

*Very small discrepancies in the totals are due to the numbers being calculated back from percentages.

It will be seen that a total of 2830 individuals represented fifteen species, the number per species ranging from two to 1427. Single species were found in from two to twelve of the lakes, but none in all thirteen. The most abundant and widely distributed species, *G. lugens*, was not found in Lac Bonan, from which only two species were collected. The number of species per lake varied from three to eleven out of the total of thirteen.

Diversities of the population, by Simpson's method, were calculated for each of the lakes and show a range from 1.4 to 2.3, while that for the total of all lakes was 3.25, reflecting the diversity of environment between the different lakes.

In Fig. 115 the total number of individuals for each species, on a log scale, has been plotted vertically against the number of lakes, out of thirteen, in which it occurred. The former is a measure of abundance, the latter of distribution. A species which is represented by only a single individual can

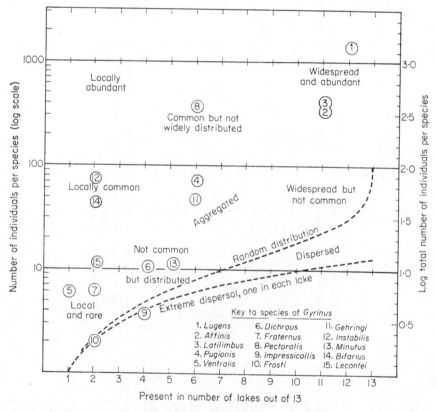

FIG. 115. Diagram illustrating relation between abundance of individuals and frequency of occurrence of Gyrinid beetles in thirteen lakes in Mont Tremblant National Park, Montreal, Canada, and differences in types of distribution in space.

naturally appear in only a single lake, but a large number of individuals may be either widely spread or concentrated in a few lakes. On the figure there has also been shown, as dotted lines, the distribution expected by purely random causes, and the distribution by extreme dispersal, each individual taking an unoccupied habitat as far as this is possible. All parts of the diagram above the line of random distribution show some measure of aggregation, and below the random distribution (to the limit of one individual for each lake) some form of dispersion. Thus the different parts of the diagram indicate different types of distribution, as, for example, "rare and local" "locally common" or "common and widespread".

It will be seen that in the present case—within the limits of error from the comparatively few data—the species group themselves into categories as follows:

Local and rare	Locally common	Wide distribution but not common	Common but not wide distribution	Widespread and abundant
ventralis	*instabilis*	*dichrous*	*pugionis*	*lugens*
fraternus	*bifarius*	*impressicollis*	*pectoralis*	*affinis*
frosti		*minutus*	*gehringi*	*latilimbus*
lecontei				

The only species showing dispersion is *G. impressicollis* with four individuals each in its own lake, but this is almost certainly not significant. The greatest aggregation is found in *G. instabilis* and *G. bifarius*.

THE ILLINOIS CHRISTMAS BIRD CENSUS

THE DISTRIBUTION OF SPECIES IN THIRTEEN LOCATIONS IN RELATION TO ABUNDANCE

In the Christmas bird census taken each year in Illinois, U.S.A., counts were made in a number of localities. In 1956 there were thirteen of these. The information for this year has been summarized in Fig. 116, which shows by the number on the diagram the name of the species according to the list in the Appendix C. The position of the number vertically is a measure, on a log scale, of the number of birds observed during the census, and horizontally of the number of locations out of thirteen in which the species was seen.

It is obvious that any small number of individuals cannot be found in a greater number of localities, and this extreme dispersion is shown by the lower dotted line in the diagram, and indicates one bird per locality. The upper dotted line represents the random distribution, and shows the number of locations expected (from the binomial expansion) to be occupied by small numbers of birds if their distribution was entirely by chance. For example, if twenty-three birds were distributed at random in thirteen locations, there would be an even chance of two locations being unoccupied. Any record below this line, if significant, would imply dispersion rather than the normal tendency for aggregation. With this comparatively small number of locations,

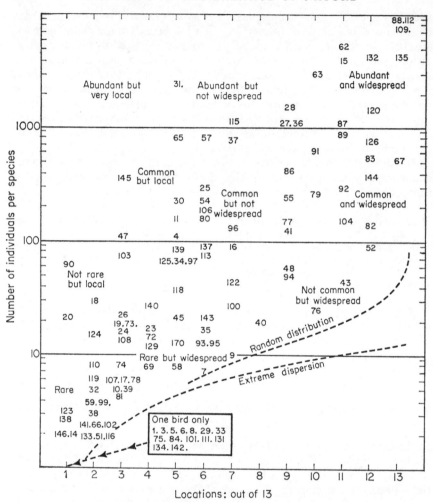

FIG. 116. Illinois Christmas Bird Census: relation between numbers of individuals and numbers of localities out of thirteen in which the birds were recorded, showing different types of distribution in space. For key to numbers see Appendix C, page 319.

which are not of similar environments, it is, however, unlikely that any record in this area of the diagram would be significant.

Turning to the diagram, we see that the different areas show different relations between abundance and distribution. The Red-Shouldered Hawk (no. 43) had, in 1956, forty-four birds which were distributed in eleven out of the thirteen localities, and this may be compared with the Carolina Chickadee (no. 90) with sixty-eight individuals all in one location. The former is not common, but widespread in many types of environment; the latter, though more common, is very restricted. The Old Squaw (no. 31) had 2414 individuals seen in 1956, but was confined to five out of the thirteen locations, much more abundant but much more restricted in range than the Carolina Chickadee. The Pied Billed Grebe (no. 7) with seven individuals in six locations, and the Great Blue Heron (no. 8) with ten birds in seven locations, show an almost random dispersion, but it is possible, that these few localities all contained the aquatic habitat that the species require.

It is not intended here to emphasize the application of the technique to particular species, except in so far as to show its value for the comparative study of ecological relationships, and the form of mathematical pattern in them.

The number of species in 1, 2, 3, etc., out of the thirteen locations in each of 4 years is given in Table 136, and diagrammatically in Fig 117. It will be seen that in each year the main frequency peak is at the lower end, with, except in 1956, a subsidiary peak at the upper end. This is shown most clearly in the smoothed curve for the average of the 4 years, and is a reversed J curve.

TABLE 136. *Illinois Bird Census: number of species which appeared in different numbers of localities out of a total of thirteen*

No. of locations out of 13	1954	1955	1956	1957	Average
1	26	24	20	13	20.75
2	14	14	14	20	15.50
3	11	13	15	11	12.50
4	10	7	6	13	9.00
5	10	5	12	2	7.25
6	8	8	12	7	8.75
7	6	6	7	3	5.50
8	8	9	1	7	6.25
9	6	4	9	4	5.75
10	5	3	4	8	5.00
11	2	4	7	5	4.50
12	3	11	7	6	6.75
13	8	7	5	11	7.75
	117	115	119	110	115.25

Although there is a general analogy between these distributions and the distribution of plant species in quadrats (see p. 68), it must be remembered that in the present, and similar cases, the areas are much larger, they are not all of the same size, and instead of being samples of a uniform environment they are usually selected to be representative of the different types of environment in the whole area under survey. In the general case of dealing with animals instead of plants we have the additional difficulty of mobility (including migration) causing almost daily changes in population structure; while in this particular case there is still another difficulty in comparing one year with another as from year to year some of the localities were replaced by others, though the majority remained unchanged throughout.

In view of these difficulties it is almost surprising to find any regular recurring pattern. The reversed J is characteristic of small samples with a high diversity which is here emphasized by the non-uniformity of the environment in the different locations.

In spite of the differences it is interesting for comparison to lay out the data in the five percentage groups suggested for plant quadrats by Raunkaier. This gives: 1–20 per cent = 43.75: 21–40 per cent = 23: 41–60 per cent = 17.5: 61–80 per cent = 13.8: 81–100 per cent = 17.2, and is shown in Fig. 117. It is somewhat similar to the distribution in quadrats for $N = 10\alpha$ and twenty quadrats as shown in Fig. 29 on p. 76.

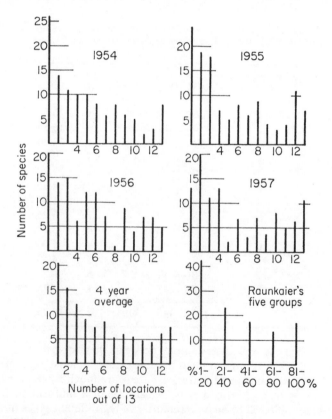

Fig. 117. Illinois Christmas Bird Census: frequency distribution of birds in different numbers of localities out of thirteen in 4 years censuses.

DISTRIBUTION OF SPECIES OF BUTTERFLIES IN DIFFERENT HABITATS AT HIGH ALTITUDES IN CALIFORNIA

Emmel and Emmel (1962) have given an account of the distribution of seventy-four species of butterflies in eleven different locations at high levels in the Donner Pass region of California. The localities were classified as follows:

I   Wet meadow: fifty-six species.

    { A. Sugar Bowl and S. Yuba River; 7260 ft; forty-four species.
    { B. Wet portion of Lodge Meadow; 7000 ft; forty-six species.

II   Dry Meadow: forty-six species.

    A. Lodge Meadow and vicinity; 7000 ft; nineteen species.
    B. Emigrant Meadow; 7500–7600 ft; thirty-two species.
    C. Summit Valley; 6800 ft; thirty-seven species.
    D. Lake Mary Road; 7000 ft; thirty-seven species.

III   Forest: 6800–7800 ft; twenty-six species.

IV   Montane; sixty species.

    A. Mt. Jubah: 8240 ft; thirty-seven species.
    B. Rock Slope leading to above; 7160–7400 ft; fifty species.
    C. Mt. Lincoln; 8383 ft; thirty-two species.
    D. Mt. Disney and Crow's Nest; 7950 ft; forty-two species.

The butterflies belonged to seven families, with the number of species in each shown in the bottom line of Table 137. It must be realized in the analysis which follows that the areas in each location were, unlike the case of quadrats, not of the same area. The exactness of the size is, however, of much less importance in the case of mobile winged insects than in plants.

Table 137 gives the distribution of the number of species which were found in different numbers of the ten* localities, in each family and for the total of all. The results are also given separately for the four major types of environment, and for the ten sublocalities. The results are shown diagrammatically in Fig. 118. It will be seen that for all the families together the distribution of species found in 1, 2, 3, etc., locations is not unlike the U-shaped curve which appears in quadrat analysis, with more species either in one, or in all of the locations, than in any of the intermediate numbers. However, for the ten localities (Fig. 118A) there is a distinct rise in the middle with species in five out of the ten localities; and with the four main divisions only, the intermediate numbers are very close to both extremes.

An examination of the data family by family, shows that in the Lycaenidae —with more species than any other—the species are much more localized.

* In Emmel's first table of distribution localities I A and B were combined.

TABLE 137. *Frequency distribution of species of butterflies in ten localities in four ecological environments in the high sierras of California*

Number of habitats out of 10	Papilionidae	Pieridae	Danaidae	Satyridae	Nymphalidae	Lycaenidae	Hesperidae	All families	Excluding Lycaenidae
				Number of species					
1	1	2	—	2	3	5	2	15	10
2	—	1	—	—	2	4	—	7	3
3	1	—	—	—	3	3	—	7	4
4	1	1	—	—	—	4	—	6	2
5	1	2	—	—	2	5	1	11	6
6	1	—	—	—	2	3	1	7	4
7	—	1	—	—	1	1	1	4	3
8	—	—	—	—	—	—	—	0	0
9	—	—	—	—	1	2	—	3	1
10	—	1	1	—	6	1	5	14	13
Habitats out of 4									
1	2	3	—	2	6	7	2	23	16
2	1	1	—	—	4	5	3	14	9
3	1	1	—	—	3	13	0	18	5
4	1	3	1	—	7	3	5	19	16
Total species in family									
	5	8	1	2	20	28	10	74	46

Only four species out of twenty-eight are in seven or more of the ten localities, and only two species out of the twenty-eight in all four environment types. The original data show that only four of the twenty-eight species of Lycaenidae occur in the "forest" environment, as compared with about 50 per cent in the rest of the families.

If the numbers for the Lycaenidae are removed from the total (last column in Table 137 and C and D in Fig. 118) the results are very much closer to the typical U distribution. There are a few more species in all locations than in one only, but considerably more in each of these than in any of the intermediate classes.

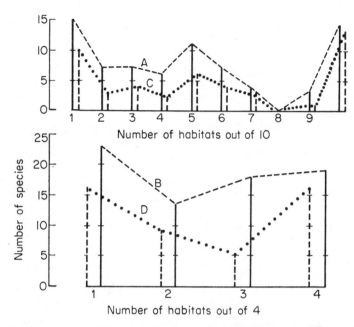

FIG. 118. Frequency distribution of species of butterflies in ten different localities in four different ecological types in the high sierras of California.

THE RELATION BETWEEN THE ABUNDANCE OF ANIMALS AND THEIR REGULARITY OF APPEARANCE AT DIFFERENT TIMES

WINTER COUNTS OF BIRDS NEAR LONDON

In the winter months of 1931 to 1937, L. Parmenter made a series of counts of the number of birds seen on "Mitcham Common, Beddington Lane and the Sewage Farm", about 8 miles S.W. of London, the details of which he kindly put at my disposal. Of these counts twenty-six were suitable for analysis, and they included a total of 159,190 birds belonging to sixty-five species. Four species occurred only once; while the most abundant species, the Starling (*Sturnis vulgaris*) was responsible for 89,989 individuals or 56.5 per cent of the total population. The number of individuals for each species, and the number of counts out of twenty-six in which the species was observed are shown in Table 138.

In Fig. 119 the total number for each species on a log scale is plotted against the number of positive counts; the former being a measure of abundance, the latter of regularity of appearance. The diagram is of a similar type to that used for the occurrence of species in different areas (see p. 274), but here the irregularity is in time and not in space.

In most species greater abundance is associated with appearance in a high

TABLE 138. *Birds observed in a series of winter observations near London, with the total number of individuals and the numbers of times out of twenty-six observations that they were seen*

Species	Individ.	Counts Ex 26	Species	Individ.	Counts Ex 26
1 Hooded crow	16	9	34 Wren	115	26
2 Carrion crow	611	26	35 Kingfisher	3	1
3 Rook	1128	22	36 Green woodpecker	6	5
4 Jackdaw	25	8	37 Gr. speck; woodpecker	14	11
5 Jay	12	8	38 Less. speck: woodpecker	1	1
6 Starling	89,989	26	39 Little owl	6	4
7 Greenfinch	417	24	40 Kestrel	34	20
8 Lesser redpoll	1	1	41 Heron	11	7
9 Goldfinch	14	6	42 Mute swan	188	26
10 Linnet	108	18	43 Mallard	433	26
11 Bullfinch	1	1	44 Gadwall	2	1
12 Chaffinch	3681	26	45 Teal	7	2
13 Brambling	238	15	46 Wigeon	10	2
14 Reed bunting	76	20	47 Pochard	269	25
15 Yellow hammer	8	6	48 Tufted duck	458	25
16 House Sparrow	5156	26	49 Little grebe	64	23
17 Tree sparrow	1710	26	50 Wood pigeon	1084	25
18 Sky lark	2152	26	51 Stock dove	131	13
19 Meadow pipit	1658	26	52 Curlew	1	1
20 Grey wagtail	114	25	53 Snipe	450	26
21 Pied wagtail	697	26	54 Redshank	4	3
22 Great tit	141	25	55 Golden plover	12	6
23 Blue tit	234	26	56 Lapwing	11,875	26
24 Long-tailed tit	6	2	57 Black-headed gull	39,124	26
25 Gold crest	4	2	58 Common gull	426	26
26 Fieldfare	920	24	59 Herring gull	660	23
27 Missel thrush	84	22	60 Lesser black-b. gull	1	1
28 Song thrush	451	26	61 Water rail	3	2
29 Redwing	2088	26	62 Moor hen	1074	26
30 Blackbird	266	26	63 Coot	1851	26
31 Stone chat	21	14	64 Partridge	16	5
32 Robin	117	26	65 Red-legged partridge	3	2
33 Hedge sparrow	84	26			

proportion of the counts. Irregularity of occurrence is, however, suggested in the Wigeon (46) with ten birds confined to two counts; the Jackdaw (4) with twenty-eight birds in eight counts; the Stock Dove (51) with 131 birds in thirteen counts; and the Brambling (13) with 238 birds in fifteen counts. In both the latter species some individuals, if distributed randomly, would have occurred in all or nearly all of the counts.

The frequency distribution of species according to the number of counts in which they were observed is shown diagrammatically in Table 139, the numbers in the table representing the species as shown in the previous Table 138. By analogy with Raunkaier's classification of species in quadrats into five groups (see p. 74) we can classify these sixty-five species of birds according to the percentage of counts in which they occur as in Table 140.

The result, with a low peak at the bottom end and a high peak at the top—a
J curve—is very typical of some of Raunkaier's distributions where there
are a large number of individuals in each group. The problem is here, however,
more complicated, as the different samples were not of the same size, actually
ranging from 2208 to 9390 birds per count.

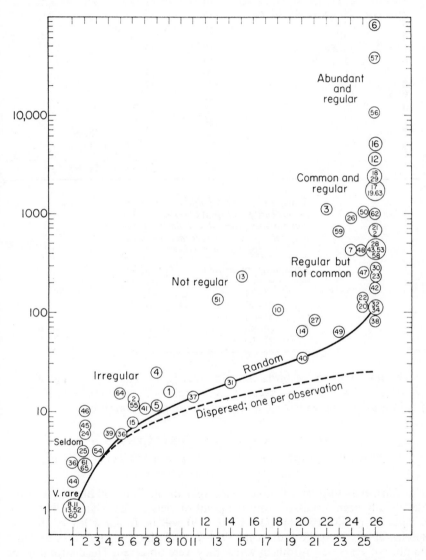

FIG. 119. Relation between abundance and regularity of appearance of winter birds
near London from records by Parmenter showing different types of distribution in
time.

TABLE 139. *Number of winter counts, out of twenty-six in which different species of birds were observed near London. The numbers in each column refer to the species given in Table 138*

Counts	Diff. species	Counts	Diff. species
1	8, 11, 35, 38, 44, 52, 60	14	31
2	24, 25, 45, 46, 61, 65	15	13
3	54	16	—
4	39	17	—
5	36, 64	18	10
6	55, 15, 9	19	—
7	41	20	14, 40
8	4, 5	21	27
9	1	22	3
10	—	23	49, 59
11	37	24	7, 26
12	—	25	20, 22, 47, 48, 50
13	51	26	2, 6, 12, 16, 17, 18, 19, 21, 23, 28, 29, 30, 32, 33, 34, 42, 43, 53, 56, 57, 58, 62, 63

TABLE 140. *Data in Table 117 arranged in percentages of total observations for comparison with Raunkaier's classification of the occurrence of plant species in quadrats*

Percentage of counts	No. of sp.
1–20	18
21–40	6
41–60	4
61–80	4
81–100	33

The total number of birds counted, 159,190 in sixty-five species, gives a diversity of 6.4, while the average per sample 6023 with 40.5 species gives 5.8, indicating a considerable uniformity among the samples.

ILLINOIS CHRISTMAS BIRD CENSUS

DISTRIBUTION OF SPECIES IN DIFFERENT YEARS AND ITS RELATION TO ABUNDANCE

The Christmas bird census taken each year in the State of Illinois, U.S.A., has already been discussed from the point of view of the relative abundance of species (p. 50). In Appendix C (p. 319) will be found a list of the 146 species observed in the four years 1954–7, with the total numbers and the number of years out of four in which they were observed. The distribution of the species, according to their abundance (in $\times 3$ log classes) and number of years, is shown in Table 141, where the number refers to the particular species in Appendix C.

TABLE 141. *Illinois Bird Census: relation of total abundance to occurrence in different numbers of years out of 4. The numbers refer to the species given in Appendix C, p. 319*

Indiv. per species	1 year only	2 years	3 years	All 4 years	Total species
XIV over 797,161				15	1
XIII 265,721–797,161					0
XII 88,574–265,720				109	1
XI 29,525–88,573				112	2
X 9842–29,524				16, 62, 88, 132, 135	5
IX 3281–9841				25, 27, 28, 31, 36, 37, 63, 87, 89, 115, 120, 126	12
VIII 1094–3280			65	11, 57, 67, 83, 86, 91, 92, 106, 144	10
VII 365–1093			116	30, 41, 54, 55, 77 79, 80, 82, 96, 100, 104, 113, 118, 119, 122, 125, 145	10
VI 122–364			121	4, 43, 47, 48, 52, 73, 90, 93, 94, 97, 103, 124, 137, 139, 143	16
V 41–121		128	1461	18, 19, 23, 24, 32, 34, 35, 40, 45, 58, 69, 70, 72, 76, 108, 140	19
IV 14–40	53, 56, 64	136	20, 74, 105, 114, 123, 127	7, 9, 39, 78, 81, 95, 107, 129, 133	19
III 5–13	21, 50, 85, 110	10	17, 38, 59, 99, 141	75	11
II 2–4	13, 14, 22, 49, 102, 138	42, 66, 71, 98, 117, 131	1, 8, 33, 51, 68, 101		18
I 1	2, 3, 5, 6, 12, 29, 44, 46, 60, 61, 84, 111, 130, 134, 142				15
Total in year group	28	9	21	89	

First it will be noticed that the distribution of the species in different numbers of years is as follows: 1 year only, twenty-eight; 2 years, nine; 3 years, twenty-one; All 4 years eighty-nine. This J-shaped distribution (see p. 76) is typical of species distributed among a small number of large samples, and indicates general regularity between the different years.

Of the species which occurred in a single year only, the most abundant was the "Prairie Chicken" with thirty-eight individuals, which could be described as not rare but uncertain or irregular in appearance. The species occurring in 2 years ranged in total abundance from two to seventy-two individuals, the latter, the White-winged Crossbill (no. 128), coming in the category of common in some years but irregular in occurrence. The species in 3 years ranged from three to 1688 individuals, the former being very rare but distributed; the latter, Bonaparte's Gull (no. 65) with 1688 individuals, and perhaps also the Rusty Blackbird with 853, would come under the category of quite common but irregular and absent in some years. The species occurring in all 4 years include the Saw-Whet Owl (no. 75) with only seven individuals, and the Pied-billed Grebe (no. 7) with fourteen. Both these could be described as regular in appearance in spite of their rarity, in contrast to the Prairie Chicken, already discussed, the thirty-eight individuals of which appeared in a single year.

In all these discussions there is, of course, the error due to differences of efficiency of different observers in different years. There is probably also a biased error due to a greater interest in the presence of the rarer species.

DISTRIBUTION OF MEALY-BUGS IN COLONIES ON COCOA TREES IN WEST AFRICA

During a visit to West Africa in April 1953 I was given information, by P. B. Cornwell of the Cocoa Research Station at Tafo, Ghana, about the number of mealy-bugs, *Pseudococcus njalensis*, Laing, in colonies on cocoa trees. The insects form small colonies on a tree and these increase in size and eventually break up to form new ones. The data for the number of colonies with different numbers of insects on three single trees are shown in Table 142 and in Fig. 120.

It will be seen that tree 121 had 2705 mealy-bugs in 179 colonies, with an arithmetic mean of nine insects per colony: tree 123 had 5498 insects in 453 colonies with a mean of 12.2; and tree 154 had 7153 insects in 426 colonies with a mean of 16.8.

An examination of the distribution shows that it is clearly not of the log-series type, so from the number of colonies on each tree in the $\times$ 3 class size Fig. 121 was made using a probability scale. It will be seen that on each tree the distribution of colonies of different sizes fits moderately close to a log normal, tree 154, which has the highest population of insects, giving the closest fit. In trees 121 and 123 there appear to be too many colonies in Class IV and too few in Class V, but as already explained, a very small change in Class V makes a considerable difference to the position on the probability

TABLE 142. *Mealy-bugs on cocoa trees in Ghana. Distribution of colonies with different numbers of individuals on three trees in which every colony was examined*

Insects per colony	121	Tree No. 123	No. of Colonies 154	Insects per colony	121	Tree No. 123	No. of Colonies 154
1	16 (16)	41 (41)	31 (31)	31	—	2	3
2	12	44	28	32	—	5	3
3	10	37	25	33	1	1	2
4	13 (35)	25 (106)	20 (73)	34	—	2	1
5	12	42	27	35	1	5	5
6	7	23	16	36	1	—	5
7	13	20	22	37	—	4	5
8	5	19	13	38	—	3	2
9	9	24	20	39	1	2	1
10	4	20	14	40	— (53)	2 (107)	— (132)
11	6	11	12	41	1	—	3
12	5	13	10	42	—	—	—
13	5 (66)	11 (183)	9 (143)	43	—	—	2
14	5	10	2	44	1	2	3
15	5	9	10	45	—	—	3
16	3	7	11	46	—	1	—
17	7	8	13	47	—	—	2
18	3	8	8	48	1	—	1
19	6	6	7	49	—	—	2
20	3	6	8	50	—	—	1
21	—	4	7	51	—	—	1
22	3	4	12	52	—	—	1
23	1	4	2	53	1	—	1
24	3	4	3	54	—	—	1
25	1	5	4	55	—	1	—
26	3	1	6	56	—	—	—
27	2	2	3	57	—	2	3
28	—	1	—	58	1	—	2
29	2	1	3	59	—	—	—
30	2	1	6	60	—	1	3

Tree 121, also at 64, 66, 130, and 256.
Tree 123, also at 62, 66, 67, 75, 94, 96, 125, 129, and 142.
Tree 154, also at 66, 71, 72, 74, 89, 92 (2), 97, 98, 101, 106, 115, 128, 132, 135, 137, 138, and 146.

scale. The axes are approximately parallel, indicating similar standard deviations of 0.42 to 0.47 on the $\log_{10}$ scale. Tree 154 has the slightly higher value, but it is doubtful if there is any significance in this.

It is interesting to note that the medians of the log normals, shown in Table 143 from the figure, are not in the same sequence as the arithmetic means. Tree 123 has the most colonies but the lowest median, and has a higher arithmetic mean than tree 121, but a lower median.

It appears from the above that the pattern of mealy-bug colonies of different sizes on a single tree is very close to a log-normal distribution.

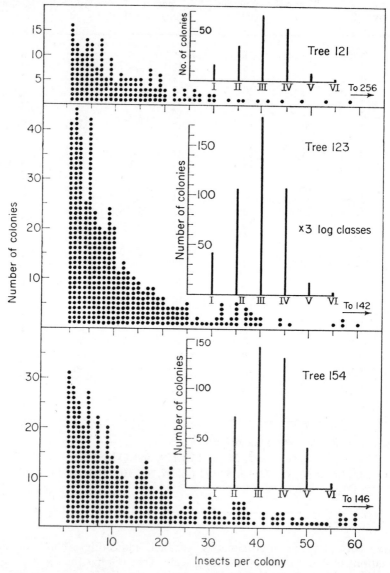

Fig. 120. Frequency distribution of mealy-bugs (Coccidae) in colonies of different sizes on three cocoa trees in Ghana, West Africa, with both arithmetic and logarithmic scales of colony size.

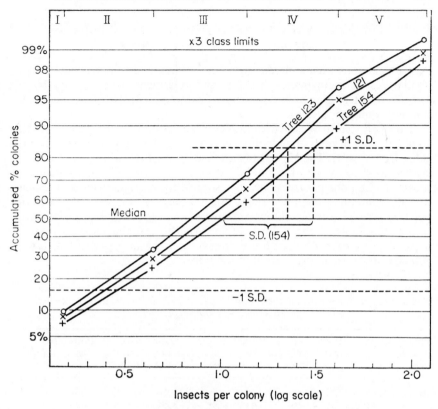

FIG. 121. Colonies of mealy-bugs in W. Africa. Data as in Fig. 120, but with pro-bability scales of percentage colonies, indicating in each tree a close fit to a log-normal distribution of colonies in different sizes.

TABLE 143. *Frequency distribution of colonies of mealy-bugs of different sizes in × 3 log classes*

Tree	Class I	II	III	IV	V	VI
121	16	35	66	53	7	2
123	41	106	183	107	13	3
154	31	73	145	132	41	6

	Total colonies	Total insects	Arith. mean	Median log normal	Standard deviation
121	179	2507	8.99	0.93 = 8.5	0.42
123	453	5498	12.14	0.85 = 7.1	0.42
54	426	7153	16.79	1.02 = 10.5	0.47

THE FREQUENCY DISTRIBUTION OF COCOA PODS ON COCOA TREES IN WEST AFRICA

A problem somewhat different from those already examined, but showing similarities in mathematical pattern and treatment, is illustrated by the frequency distribution of fruits on individual trees. It is also often of considerable economic importance.

During a visit to West Africa in 1953 I was given two sets of data relating to the number of pods on individual cocoa trees, one set from Nigeria and the other from Ghana. The Nigerian figures (see Table 144 and Figs. 122 and 123) related to 4936 trees with a total yield of 73,123 pods, giving an arithmetic mean of 14.8 pods per tree. The highest yield for a single tree was 172 pods. The distribution is definitely skew, and the median was just over ten pods per tree. We have thus another example of the typical biological frequency distribution in which the majority of units (in this case the pods) are found in a small proportion of the groups (trees), and the majority of the groups provide only a small proportion of the units.

TABLE 144. *Frequency distribution of 4936 cocoa trees in Nigeria, West Africa, with different numbers of pods per tree*

Pods per tree	No. of trees	Pods per tree	No. of trees	Pods per tree	No. of trees	Pods per tree	No. of trees	Pods per tree	No. of trees
0	104	20	78	40	23 (1711)	60	2	80	1
1	392 (392)	21	86	41	30	61	4	82	1
2	293	22	84	42	18	62	7	84	1
3	274	23	66	43	16	63	3	85	1
4	244 (811)	24	69	44	13	64	5	86	2
5	198	25	56	45	11	65	2	87	1
6	232	26	56	46	22	66	2	88	1
7	215	27	52	47	17	67	5	91	1
8	185	28	52	48	10	68	3	93	1
9	180	29	57	49	11	69	3	101	1
10	170	30	36	50	15	70	6	103	1
11	157	31	46	51	15	71	4	105	1
12	150	32	39	52	4	72	—	106	1
13	120 (1607)	33	32	53	6	73	3	108	2
14	135	34	33	54	9	74	2	112	1 (306)
15	132	35	29	55	12	75	1	122	1
16	117	36	37	56	12	76	—	126	1
17	113	37	32	57	6	77	—	148	1
18	105	38	26	58	5	78	—	159	1
19	93	39	27	59	4	79	1	172	1 (5)

Total trees = 4936.   Total pods = 73,123

Arithmetic mean pods per tree = 14.8

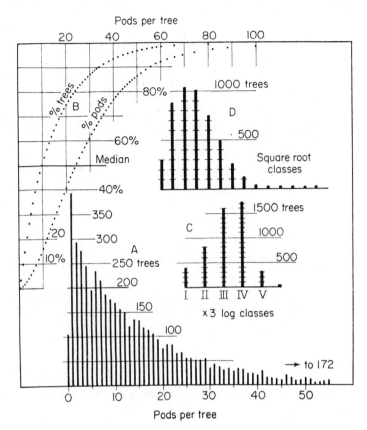

FIG. 122. Frequency distribution of cocoa trees with different numbers of pods in Nigeria, West Africa. A, on arithmetic scale of numbers of pods; B, showing accumulated percentage of total pods and total trees with increasing crop per tree; C, pods per tree in × 3 classes, and D, the same in square-root classes.

Figure 122B, which shows the accumulative percentage of the total pods and of total trees with up to 1, 2, 3, etc., pods per tree, indicates that the 50 per cent of the trees having up to ten pods per tree provide only 15 per cent of the crop; and the 74 per cent of all the trees with up to twenty pods per tree still provide only 40 per cent of the crop. Starting from the upper end, we find that the 10 per cent of the trees having thirty-five pods or more per tree produce 30 per cent of the crop; and that half the total crop is produced by the 20 per cent of the trees with over twenty-five pods per tree. Thus a very high proportion of the trees is probably economically unsound, and it is for the planter to see if this is due to differences in age, treatment, local environment or to genetic mixture.

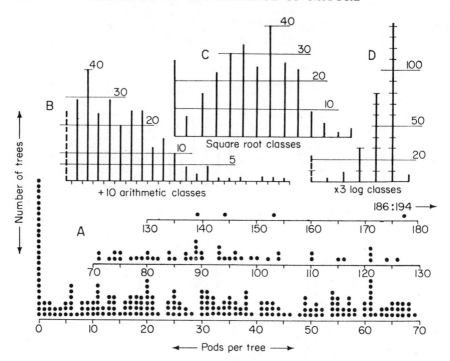

FIG. 123. Frequency distribution of cocoa trees with different numbers of pods in Ghana, West Africa. A shows pods per tree on an arithmetic scale of integers; B, on an arithmetic scale in groups of 10; C, on a square-root transformation, and D, in × 3 log classes.

The frequency distributions were sorted into × 3 log, and also into square-root classes, as shown diagrammatically in Fig. 122c and D, and also on a probability scale in Fig. 124. It will be seen that neither transformation gives a symmetrical result, the × 3 classes giving a slight overcorrection and the square root slightly under.

An attempt to fit a negative binomial to the total pods, the total trees, and the number of trees without pods gave $k = 1.7$ and $q = 9.714$, but the series calculated from these bore no relation to the observed distribution.

The data available from Ghana consisted of the yields of trees on four quarter-acre plots, on which there had been a better selection of trees of the same ages, and much better agricultural conditions. A summary of the results is given in Table 145. It will be seen that the overall arithmetic mean yield, at 43.6 pods per tree, is very much higher than with the less selected trees in Nigeria; but the highest yield per tree in the Ghana plots, at 194 pods, was not very much higher than the 172 for Nigeria. The high average yield was thus due to a great reduction in the number of the poorer trees.

The frequency distribution of all the plots is shown in Fig. 123A, from

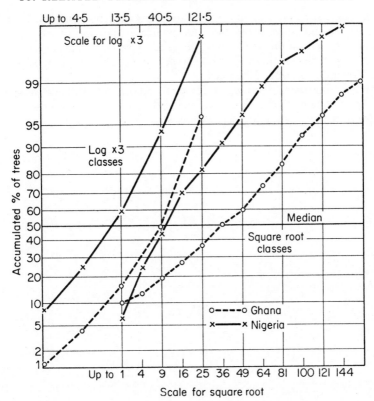

FIG. 124. Data on cocoa trees in West Africa with different numbers of pods as in Figs. 122 and 123, on "probability" diagrams, for square-root and logarithmic scales, showing absence of significant fit to the log-normal or square-root-normal distribution.

TABLE 145. *Summary of yields of pods per cocoa tree in four quarter-acre plots in Ghana, West Africa*

Plot	No. of trees	Total pods	Pods per tree		% of pods for 50% poorest trees	% of best trees giving 50% pods
			Arith. mean	Median		
A	64	3775	59.0	59	25.0	31
B	65	2258	34.7	28	19.3	23
C	46	1335	27.0	25	11.7	22
D	94	4070	43.3	32	16.2	30
All four plots						
	269	11,438				
	+2*	+377*				
	271	11,815	43.0	35	17.5	22.3

*Two high-yielding trees were unfortunately not allocated to their particular plot.

which this reduction of the poorer yielding trees can be well seen. Even here, however, as shown in Table 145, the greater proportion of trees is of low production, and on the whole four plots the 50 per cent poorest trees produce only 17.5 per cent of the crop, while 50 per cent of the total crop is produced by just over 22 per cent of the trees.

An examination of the Ghana figures by class transformation, as shown in Fig. 123 indicates that a more symmetrical distribution is produced by the square root than by the log classes.

Chapter 11

FINAL SURVEY AND COMMENTS

In the study of living creatures—both animal and plant—a large number of problems can be approached from the mathematical point of view by considering the frequency distribution of the number of groups with different numbers of units. This present survey has included the frequency of species with different numbers of individuals: of genera with different numbers of species; of hosts with different numbers of parasites; of species appearing in different numbers of localities or at different times of observation; and many other similar problems.

Such balances of abundance are, of course, only momentary, and their composition is always changing. But it is of importance to ask if, while the sizes of the different groups making up the particular population under discussion are changing, the basic pattern in which they are arranged may be much more constant, and so a much more important characteristic of the balance than the position within it of any particular group or unit.

This study, therefore, has been based on an attempt to get "static" cross-sections of mixed wild populations at intervals in the same population, or in different populations, in order to see if the dynamic changes which are always going on determine the static pattern or merely ring changes on the same pattern.

One of the first results that emerged was the almost universal prevalence in population numbers of geometric or logarithmic changes in abundance (units per group). With animal populations, consisting of individuals grouped into species, it is easy to see that the favourable conditions which would allow a population of 100 animals to increase to 110 would allow a population of 200 to increase to 220. It is not the arithmetical addition of ten or twenty individuals which is fundamental, but the increase of 10 per cent.

Since many statistical formulae are based on the tacit assumption that variation about the mean is equal and opposite on an arithmetic scale, serious trouble can be caused by using such formula for data which vary geometrically, without a transformation of scale. For example, the application of the usual formula for a "missing plot", when applied to data on insect catches which were varying geometrically, provided the absurd result that a trap should have been expected to catch a negative number of insects (Williams, 1953 b). As a result of many years' experience I have found that the arithmetic scale of variation in numbers is the exception rather than the rule in population studies.

When a geometrically varying factor is transferred from an arithmetic to a logarithmic scale, the geometric variations become arithmetic. A multiplication by 2, at any level, becomes the addition of 0.3. Thus formulae based on arithmetic variations become applicable *after* the transformation of such data to a log scale.

It follows from the above that the geometric mean (or the mean log) is often a much better standard for comparison than the arithmetic mean.

Throughout the previous chapters it will have been seen that two frequency distributions, the log normal and the log series, appear to be most frequently applicable to the distributions found in population data.

Although the log normal is more complicated mathematically, its applicability is easy to understand, as it is the necessary result of the normal distribution being applied to populations which vary geometrically. In other words the very large number of factors, acting sometimes favourably and sometimes unfavourably, each produces, not additions and subtractions to the number of units in each group, but multiplications and divisions. Thus a logarithmic transformation of the units per group results in the normal distribution becoming applicable.

It is still uncertain whether the close fit of the log series to many frequency distributions, especially when the number of units per group is low, is a fundamental departure from the log normal, or whether it is due to the small samples from a log normal differing so little from a log series that the data may not be critical enough to distinguish. There are, however, some cases, particularly among parasites and hosts, where the log series given is a definitely better fit to the observed data than the log normal.

Fisher's original derivation of the logarithmic series was based on the assumption that the abundance of different species is distributed as a constant $\times \chi^2$. He also showed that a sample taken from a population distributed according to a log series, would also itself show a log-series distribution.

The possible origin of a log-series distribution in the case of genera and species has been discussed by Kendall (1948) and Skellam (1951). Kendall considered three evolutionary factors:

(1) the probability of a species splitting into two;
(2) the probability of a new species arriving from outside;
(3) the probability of a species disappearing.

He showed that if, at some particular time, we started with a series of genera each with one species, their distribution after a time t would be a geometric series with an abnormal zero class. Since different genera start at different times, an integration must be made which leads to a log-series distribution.

A point of considerable interest in the biological interpretation of the logarithmic series arose from a study that I made on the application of this series to the frequency distribution of spells of wet and of dry weather lasting for different numbers of days (Williams, 1952).

It had previously been suggested that these spells had a distribution in the form of a geometric series, which has the property that wherever you start in the series the remainder has the same form as the original. Thus, in a geometric series, if of all the groups p % have more than one unit, then of all those with at least two units p % will have at least three; and of all groups with 10 or more units p % will have at least 11.

In the case of the wet and fine spells it was found that the chance of a spell lasting one further day increased as the previous duration of the spell increased. This increase is a property of the logarithmic series, and is dependent on the value of "x" in the formula. Thus if $x = 0.5$ (i.e. an average of 1.44 days per spell) then 28 % of the first days will be followed by a second; 36 % of the second days will be followed by a third; and 40 % of these will be followed by a fourth; and so on. If $x = 0.9$ (average length of spells 3.9 days) then these percentage values become 61, 71.1 and 75.8 respectively.

Thus biologically we could interpret the log series in words as "nothing succeeds like success". The higher level of abundance that a species has reached, the easier will it be for it to take one step higher. The more parasites that have reached one particular host, the more likely is it that still another will find its way. At which point the increase in numbers becomes a drawback and not an advantage may well determine whether the log-series distribution continues or changes to some other basic pattern producing a sigmoid relation between abundance and total species at that level.

The range of average level of units per group—whether taken as the arithmetic or geometric mean—varies greatly from problem to problem. In some of the parasite-host relations the average is less than one parasite per host. In others (the malaria parasite, for example) it may run into millions. It has, however, been pointed out that in all such cases where, perforce, the sampling is made by the selection of complete groups (hosts), no increase in the size of the sample alters the average number of units per group. In all cases where the samples are taken by selecting of units a larger sample gives a higher number of units per group.

If the population remains for a long period without interference from outside, and if the physical environment also remains constant, a steady state of balance might theoretically be achieved. Any sudden change, such as the introduction of a new species, produces a state of "imbalance" and the various forces of population control will set about returning to a new balance, although this may be only a new variation of the older pattern. Such cases are well known in the accidental introduction of insects into a new country, where they have rapidly increased in numbers to become a major pest. Man then makes efforts to restore the previous balance by the introduction of enemies of the new pest.

Apart from these exceptional disturbances, there are many quite regular upsets in population balance of a similar type. On the physical side of the environment there are the seasonal changes of temperature and rainfall in all temperate and most tropical countries. This makes conditions easier at one time of the year than at another, varying for different species and so producing forces tending towards a different balance. The life histories of species which live for many years, or have only one brood per year, are usually adapted to fit in with, and reduce the effect of, these changes. Species with several broods in a year may find conditions favourable for a build-up of population at one season and less favourable at another. The result will be a change of their position in the population make-up at different seasons, but this, again, will probably not affect the basic pattern of balance. The overlapping of generations also tends to stabilize the population.

A more sudden form of population change, which often makes sampling difficult to interpret, is periodic or irregular migration. A population of a large number of resident species in an area, which might, in the course of time, set up a more stable balance, is suddenly affected—perhaps twice a year—by an influx or departure of great numbers of migrants, which may at times equal or exceed the resident population. The discussion above on the Mallard Duck (p. 63) in winter in Illinois is a case in point, as in four successive winters this species varied in numbers from 5.6 to 86 per cent of the total bird population.

The numbers of an immigrant species depend, of course, not on the physical condition and biological competition in the area where they congregate, but on environmental factors elsewhere, perhaps a thousand or more miles away.

Thus the study of balance and pattern in animal populations is made more complex by their ability to move, and particularly by mass migration.

In the plant world, where the complication of movement does not normally occur, there is the interesting suggestion from the work of Pidgeon and Ashby (p. 84) in Australia that the balance among annual plants and that among perennial plants in an area may be problems separate from each other, and from the balance of the population as a whole. In some ways the perennial plants can be considered as part of the environment in which the annuals have to live and die. This would be a most interesting line for further study, and has its parallel in the suggestions by Bond (p. 88) that agricultural crops, controlled and protected by man, might be considered as part of the environment for the population of "weeds" infesting the same area.

The attempt to find mathematical formulae that will fit the observed frequency distributions led to the possibility of measuring the "diversity" of a population, which is high when the number of groups is high compared with the number of units, and low when the number of groups is low. This diversity is a property of the population as a whole, and not of any of the units or groups included in it.

Diversity is independent of the basic pattern of balance and is mathematically usually one of the "constants" which are included in the general formula. The same pattern can exist with many different diversities. A difference in diversity between two populations, or between samples from them, indicates a difference in the make-up of the populations. On the other hand, identity of diversity does not necessarily imply identity of the two populations. Absolute identity of populations can only be demonstrated if samples from each have the same diversity, and also the diversity remains the same when the two samples are combined into one.

A survey of the relations between individuals and species shows that, for similar groups of animals and plants, specific diversity is higher in tropical areas and lower in cold temperate regions. It is higher in summer than in winter in the same area; and it is low in environments where there is some particularly severe limiting factor in the physical environment, as, for example, extreme dryness, great altitude or depth, an acid soil, etc.

It has therefore been suggested that the Index of Diversity can be an indication of the relative importance of the factors that are affecting the population balance as a whole. A low index would suggest the overriding importance of physical difficulties. A high diversity suggests extreme biological competition in otherwise favourable conditions.

It is important to note that with low diversities the number of individuals need not be small. This is well shown in some subarctic areas where Diptera exist in gigantic numbers, but represent very few species.

There is some evidence that Generic Diversity shows similar geographic differences, being higher in tropical conditions and lower in cold temperate zones. The changes in generic diversity at different seasons have also been examined, but the results are at present inconclusive.

The study of our human efforts at classifying animals and plants into species, genera and higher groups also suggests a regular mathematical form in all such schemes. This is either an indication of pure chance, or expresses some real biological relationships in the evolutionary history of species and higher groups. The close resemblance between the pattern of the individual–species and the species–genus relation is an argument in favour of some real biological meaning.

The criterion of diversity can also be applied here. It appears that different taxonomists, with their different basic ideas on what differences should justify a "genus" or other group, produce classifications with different diversities. The diversity is low for "lumpers" and high for "splitters", but the pattern behind is the same for all. Thus it seems possible that good taxonomists may be splitters or lumpers, each interpreting equally correctly the fundamental pattern of evolutionary relationships. The differences in diversity are properties of the minds of the workers. As a large number of scientists without any difference of opinion would be disastrous, we must welcome the different approaches, provided that in each case the principles have been applied with consistency.

Since all natural populations show such strong resemblances in their pattern, it would be an interesting experiment to plant, on as uniform an area as possible, a series of plots with different mixtures of seeds of grass and other pasture plants: some with only one species, some with equal mixtures of two, and so on. Then, at regular intervals of a year or less, careful surveys should be made to see how the change from human artificiality to natural balance takes place, and how long it would take for all the plots to become similar.

Something of the kind, but in a reverse direction, has been carried out at

Rothamsted Experimental Station, in the "Park Grass Plots". Here, an area of moderately uniform grassland was many years ago divided into plots, each subjected for many years to a special type of manurial treatment. A series of surveys (at too long intervals) has shown the gradual increase in individuality in the different plots. So far as I am aware the reverse experiment suggested above has not been carried out. Must we wait for another John Lawes?

It has been seen that, except in the case of sudden overwhelming immigration, no samples from natural populations contain a number of dominant species all equally abundant or one species completely outnumbering the rest. Normally the most abundant species contains 10 to 40 per cent of the total population. Almost complete dominance of a single species is found only when human interference has taken place, as in a forest plantation or a field of wheat.

I once heard it stated, as an argument in probability, that if in a strange land one found four trees of one species in a straight line, it was evidence in favour of the presence of man. If, on measurement, the trees were found to be equidistant, then the assumption became a certainty!

Perhaps, who knows, in the not-too-distant future it may be possible for space explorers to infer the presence or absence of ruling minds on planets or stars, by observations on the relative abundance or distance apart of any organisms they discover.

REFERENCES

A small number of references, consulted but not actually quoted in the text, have been added as they have a special bearing on the problems discussed.

Agrell, L. (1941). Zur oecology der Collembola. *Opusc. ent. Suppl.* 3.

Aitchison, J., and Brown, J. A. C. (1957). "The Log normal Distribution", Cambridge University Press, London.

Allan, R. M. (1956). A study of the populations of the rabbit flea, *Spilopsyllus cuniculi* (Dale) on the wild rabbit *Orictolagus cuniculus* in north-east Scotland. *Proc. R. ent. Soc. Lond.* A **31**, 145–52.

Angelier, E. (1953). L'indice de diversité de C. B. Williams et son interêt en biogeographie. *C. R. (Som) Soc. Biogeogr. Paris*, 258.

Anscombe, F. J. (1949). The statistical analysis of insect counts based on the negative binomial distribution. *Biometrics* **5**, 165–73.

Anscombe, F. J. (1950). Sampling theory of the negative binomial and logarithmic series distributions. *Biometrika* **37**, 358–82.

Archibald, E. E. A. (1949). The specific characters of plant communities. Part II, A quantitive approach. *J. Ecol.* **37**, 274–88.

Arrhenius, O. (1923). Statistical investigations on the constitution of plant associations. *Ecology* **4**, 68–73.

Ashby, E. (1935). The quantitative analysis of vegetation. *Ann. Bot., Lond.* **49**, 779–802.

Bagenal, T. B. (1951). A note on the papers of Elton and Williams on the generic relations of species in small ecological communities. *J. Anim. Ecol.* **20**, 242–5.

Barnes, H. F. (1932). Studies of fluctuations in insect populations. I, Infestation of Broadbalk wheat by the wheat-blossom midge. *J. Anim. Ecol.* **1**, 12–31.

Beare, T. H., and Donisthorpe, H. StJ. K. (1904). "Catalogue of the British Coleoptera", London.

Bedwell, E. C. (1945). The county distribution of the British Hemiptera-Heteroptera. *Ent. mon. Mag.* **81**, 253–73.

Bentham, G., and Hooker, J. D. (1904, 1912). "Handbook of the British Fauna", London (two editions).

Bertram, D. S. (1949). Studies in the transmission of Cotton Rat Filariasis. I, The variabilities of the intensities of infection of the individuals of the vector *Liponyssus bacoti*. *Ann. trop. Med. Parasit.* **43**, 313–32.

Best, A. E. G. (1950). Records of light traps, 1947–48. *Ent. Gaz.* **1**, 228–9.

Blackman, G. E. (1935). A study by statistical methods of the distribution of species in grassland associations. *Ann. Bot., Lond.* **49**, 749–77.

Bliss, C. I. (1958). The analysis of insect counts as negative binomial distributions *Proc. X Int. Long. Ent.* (1956) **2**, 1015–32.

Bond, T. E. T. (1947). Some Ceylon examples of the logarithmic series and the index of diversity of plant and animal populations. *Ceylon J. Sci.* A. **12**, 195–202.

Bond, T. E. T. (1952). Applicability of the logarithmic series to the distribution of the British Heiracea and other plants. *Proc. Linn. Soc. Lond.* **163**, 29–38.

Brenchley, W. E. (1924). "Manuring of Grass Land for Hay", London.
Brenchley, W. E., and Adam, H. (1915). Recolonisation of cultivated land allowed to revert to natural conditions. *J. Ecol.* 3, 193–210.
Brereton, J. le G. (1957). Defoliation in rain forest. *Aust. J. Sci.* 19, 204–5.
Brian, M. V. (1953). Species frequencies in random samples of animal populations. *J. Anim. Ecol.* 22, 57–64.
British Ornithological Union (1952). "Check List of the Birds of Great Britain and Ireland", London.
Buxton, P. A. (1936–41). Studies on populations of human head-lice (Pediculus humanus capitis) I–IV. *Parasitology* 28, 92–97; 30, 85–110; 32, 296–302; and 33, 224–42.

Cailleux, A. (1952). Richesse des flores; physiologie et évolution. Typescript presented to colloquium on "Evolution" in Paris.
Cailleux, A. (1953). "Biogeographie Mondiale", Presse Universitaires, Paris.
Cain, S. A. (1938). The species-area curve. *Amer. Midl. Nat.* 19, 573–81.
Cassie, R. M. (1962). Frequency distribution models in the ecology of plankton and other organisms. *J. Anim. Ecol.* 31, 65–92.
Chamberlin, J. C. (1924). Concerning the hollow curve of distribution. *Amer. Nat.* 58, 350–74.
Chapman, R. N. (1931). "Animal Ecology, with Special Reference to Insects", McGraw-Hill, New York.
Chapman, V. J. (1934). Appendix II, "Floral List", pp. 229–31, *in* A. Steers, "Scolt Head Island", Cambridge University Press, London.
China, W. E. (1943). The generic names of the British Hemiptera-Heteroptera. "Generic Names of British Insects", Part 8, Royal Entomological Society, London.
Christophers, S. R. (1924). The mechanism of immunity against malaria in communities living under hyper-endemic conditions. *Indian J. med. Res.* 12, 273–94.
Clapham, A. R., Tutin, T. G., and Warburg, E. F. (1952). "Flora of the British Isles", Cambridge University Press, London.
Cole, L. C. (1946). A theory for analysing contagiously distributed populations. *Ecology* 27, 329–41.
Cole, L. C. (1949). The measurement of inter-specific association. *Ecology* 30, 411–24.
Corbet, A. S. (1942). The distribution of butterflies in the Malay Peninsula. *Proc. R. ent. Soc. Lond.* A. 16, 101–16.
Corbet, A. S. (1943). See under Fisher, Corbet, and Williams (1943).
Corbet, G. B. (1956). The life history and host relations of a hippoboscid fly, *Ornithomyia fringillina* Curtis. *J. Anim. Ecol.* 25, 403–420.
da Costa Lima, A., and Hathaway, C. R. (1946). Pulgas, Bibliographia Catalogo e Hospedadores. *Monogr. Inst. Osw. Cruz.* 4, 75–415.
Crighton, M. I. (1960). A study of captures of Trichoptera in a light-trap near Reading, Berkshire. *Trans. R. ent. Soc. Lond.* 112, 319–44.

Dehalu, M., and Leclercq, J. (1951). Application des series logarithmiques de Fisher-Williams à le classification des Hymenopteres Crabroniens. *Ann. Soc. R. Zool. Belg.* 82, 67–82.
Den Boer, P. J. (1958). Activiteitsperioden van loopkevers in Meijendal. *Ent Ber., Amst.* 18, 80–89.
Dirks, C. O. (1937). Biological studies in Maine moths by light-trap methods. *Bull. Me agric. Exp. Sta.* 389, 31–162.

Donisthorpe, H. StJ. K. (1939). "A Preliminary List of the Coleoptera of Windsor Forest", London.

Dunn, E. R. (1949). Relative abundance of some Panamanian snakes. *Ecology* 30, 39–57.

Dunn, E. R., and Allendoerfer, C. B. (1949). The application of Fisher's formula to collections of Panamanian snakes. *Ecology* 30, 533–6.

Easton, A. M. (1947). The Coleoptera of flood refuse: a comparison of samples from Surrey and Oxfordshire. *Ent. mon. Mag.* 83, 113–15.

Elton, C. (1946). Competition and the structure of ecological communities. *J. Anim. Ecol.* 15, 54–68.

Emmel, T. C., and Emmel, J. F. (1962). Ecological studies of Rhopalocera in a high sierran community—Donner Pass, California I, Butterfly associations and distributional factors. *J. Lep. Soc.* 16, 23–44. Also p. 136.

Ennion, E. (1960). "The House on the Shore", Routledge and Kegan Paul, London.

Evans, F. C. (1950). Relative abundance of species and the pyramid of numbers. *Ecology* 31, 631–2.

Evans, F. C., and Freeman, R. B. (1950). On the relationships of some mammal fleas to their hosts. *Ann. ent. Soc. Amer.* 43, 320–33.

Exell, A. W. (1944). Catalogue of the vascular plants of S. Tomé. British Museum, London.

Exell, A. W. (1947). Discussion on the percentage relationship calculated by Dr Williams. *Proc. Linn. Soc. Lond.* 1945–46, 108–10.

Farren, W. (1936). A list of the Lepidoptera of Wicken and the neighbouring fens, *in* J. S. Gardiner, "The Natural History of Wicken Fen", Part III, pp. 258–66, Cambridge University Press, London.

Fisher, J. (1952). Bird numbers; a discussion on the breeding populations of inland birds of England and Wales. *S. East. Nat.* 57, 1–10.

Fisher, R. A. (1941). The negative binomial distribution. *Ann. Eugen., Lond.* 11, 182–7.

Fisher, R. A., Corbet, A. S., and Williams, C. B. (1943). The relation between the number of species and the number of individuals in a random sample of an animal population. *J. Anim. Ecol.* 12, 42–58.

Foster, A. H. (1937). A list of the Lepidoptera of Hertfordshire. *Trans. Herts. nat. Hist. Soc.* 20, 171–279.

Freeman, J. A. (1946). The distribution of spiders and mites up to 300 ft in the air. *J. Anim. Ecol.* 15, 69–74.

Fryer, G. (1956). A report on the parasitic Copepoda and Branchiura of the fishes of Lake Nyasa. *Proc. zool. Soc. Lond.* 127, 293–344.

Fryer, G. (1959). A report on the parasitic Copepoda and Branchiura of the fishes of Lake Bangweulu, Northern Rhodesia. *Proc. zool. Soc. Lond.* 132, 517–50.

Fryer, G. (1961). The parasitic Copepoda and Branchiura of the fishes of Lake Victoria and the Victorian Nile. *Proc. zool. Soc. Lond.* 137, 41–60.

Gaddum, J. H. (1945). Log-normal distributions. *Nature, Lond.* 156, 463–6 and 747.

Garthside, S. (1928). Quantitative studies upon the insect fauna of Jack Pine environments. Unpublished thesis in the library of the University of Minnesota.

Gause, G. F. (1934). "The Struggle for Existence", Baltimore.

Gleason, H. A. (1922). On the relation between species and area. *Ecology* 3, 158–62.

Gleason, H. A. (1925). Species and area. *Ecology* 6, 66–74.

Gower, J. C. (1961). A note on some asymptotic properties of the logarithmic-series distribution. *Biometrika* 48, 212–15.

Graham, S. A. (1933). The influence of civilisation on the insect fauna of forests. *Ann. ent. Soc. Amer.* **26**, 497–503.

Grundy, P. M. (1951). The expected frequencies in a sample of an animal population in which the abundances of species are log-normally distributed. Part I, *Biometrika* **37**, 427–34.

Hafez, M. (1953). Studies on Tachina larvarum L. *Bull. Soc. Fouad. Ent.* **37**, 305–25.

Harrison, J. L. (1960). Faunal diversity in Australian tropical rain forest. *Aust. J. Sci.* **22**, 424–5.

Herdan, G. (1957). The mathematical relation between the numbers of diseases and the numbers of patients in a community. *J. R. statist. Soc.* A **120**, 320–30.

Hopkins, B. (1955). Species area relations of plant communities. *J. Ecol.* **43**, 409–26.

Houtman, G. (1961). Vlindervangsten te Hoorn. *Ent. Ber., Amst.* **21**, 177–80.

Hunter, G. C., and Quenouille, M. H. (1952). A statistical examination of the worm egg count sampling technique for sheep. *J. Helminth.* **26**, 157–70.

Hyde, H. A., and Williams, D. A. (1949). A census of mould spores in the atmosphere. *Nature, Lond.* **164**, 668–9.

Illinois Christmas Bird Census, 1954–1957, in *Audubon (annu.) Bull.*
  (1954). (P. H. Lobik). *Bull.* **93**, 6–12.
  (1955). (P. H. Lobik). *Bull.* **97**, 6–12.
  (1956). P. H. Lobik. *Bull.* **101**, 8–9.
  (1957). P. H. Lobik. *Bull.* **105**, 5–13.

Jaccard, P. (1902). Gesetze der pflanzenvertheilung in der alpinen region. *Flora, Jena* **90**, 349–77.

Jaccard, P. (1902 *a*). Lois de distribution florale dans la zone alpine. *Bull. Soc. Vaud. Sci. Nat.* **38**, 69–130.

Jaccard, P. (1908). Nouvelles recherches sur la distribution florale. *Bull. Soc. Vaud. Sci. Nat.* **44**, 223–70.

Jaccard, P. (1912). The distribution of the flora in the alpine zone. *New Phytol.* **11**, 37–50.

Jaccard, P. (1914). Etude comparative de la distribution florale dans quelques formations terrestres et aquatiques. *Rev. Gen Bot. Paris.* **26**, 5–21 and 49–78.

Jaccard, P. (1920). Une exception apparente à la loi du coefficient generique. *Bull. Soc. Vaud. Sci. Nat.* **53**, proc. verb. 74–76.

Jaccard, P. (1928). Phytosociologie et Phytodémographie. *Bull. Soc. Vaud. Sci. Nat.* **61**, 441–63.

Jaccard, P. (1928 *a*). Die statisch-floristische Methode als Grundlage der Pflanzensociologie. *Handb. biol. ArbMeth.* **11**, 165–202.

Jaccard, P. (1939). Un cas particular concernant le coefficient generique. *Bull. Soc. Vaud. Sci. Nat.* **60**, 249–53.

Jaccard, P. (1941). Sur le coefficient generique. *Chron. Bot.* **6**, 361–4 and 389–91.

Jenkins, J. T. (1936). "Fishes of the British Isles" (2nd Edition), Warne and Co., London.

Johnson, C. G., and Mellanby, K. (1942). The parasitology of human scabies. *Parasitology* **34**, 285–90.

Jones, E. W. (1945). The index of diversity as applied to ecological problems. *Nature, Lond.* **155**, 390.

Kendall, D. G. (1948). On some modes of population growth leading to R. A. Fisher's logarithmic series distribution. *Biometrika* **35**, 6–15.

Killington, F. J. (1937). "Generic Names of the British Neuroptera", Royal Entomological Society, London.

Kirby, W. (1904). "A Synonymic Catalogue of the Orthoptera", Vol. I, British Museum, London.

Kleczkowski, I. (1949). Transformation of lesion counts for statistical analysis. *Ann. appl. Biol.* **36**, 139–55.

Kontkanen, P. (1950). Quantitative and seasonal studies on the leafhopper fauna of the field stratum in open areas in North Karelia. *Ann. (Zool.) Soc. Zool. Bot. Fen. Vanama*, **13**, 1–91.

Lane, C., and Rothschild, M. (1961). Notes on migrant Lepidoptera captured in Israel. 2, Moths captured on the sea shore at Herzlia. *Entomologist* **94**, 295–304.

Laurence, B. R. (1955). The ecology of some British Sphaeroceridae (Boboridae, Diptera). *J. Anim. Ecol.* **24**, 187–99.

Leclercq, J. (1954). Monographie systematique, phylogenetique et zoogeographique des Hymenopteres Crabroniens. *Thèse Fac. Sci. Univ. Liége.*

MacArthur, R. H. (1957). On the relative abundance of bird species. *Proc. nat. Acad. Sci., Wash.* **43**, 293–5.

McGregor, E. A. (1924). Painted Lady Butterfly (*Vanessa cardui*). *Insect Pest Surv. Bull. U.S.* **4**, 70.

Margalef, R. (1949 a). Les asociaciones de algas en las aquas dulces de pequeño volumen del nor-este de España. *Vegetatio* **1**, 258–84.

Margalef, R. (1949 b). Una aplicación de las serias logarítmicas a la fitosociologia. *Publ. Inst. Biol. Appl. Barcelona* **6**, 59–72.

Margalef, R. (1956 a). Información y diversidad especifica en las comunidades de organismus. *Inv. Pesq.* **3**, 99–106.

Margalef, R. (1956 b). "La Diversidad de Especies en las Poblaciones Mixtas Naturales y en el Estudio del Dinamismo de las Mismas", 229–43 (Tomo Homen. Posth. Dr F. Pardillo), Universidad de Barcelona.

Margalef, R. (1957). La teoria de la información en ecologia. *Mem. R. Acad. Barcelona.* 3rd ser. **32**, 373–449.

Marquand, E. D. (1901). "Flora of Guernsey", London.

Metcalf, Z. P. (1940). How many insects are there in the world? *Ent. News* **51**, 219–22.

Meyrick, E. (1895). "Handbook of the British Lepidoptera", London.

Meyrick, E. (1927). "Revised Handbook of the British Lepidoptera", London.

Murray, J. (1895). Summary of scientific results obtained at soundings, dredgings and trawling stations of H.M.S. *Challenger. Challenger Repts. Summary*, Part 2, 1430

Moreau, R. E. (1948). Ecological isolation in a rich tropical avi-fauna. *J. Anim. Ecol.* **17**, 113–26.

Oliver, F. W. (1943). *Proc. R. ent. Soc. Lond.* A. **18**, 87–88.

Ormiston, W. (1917). "Butterflies of Ceylon", Colombo.

Palmén, E. (1944). Die anemohydrochore Ausbreitung der Insekten als zoogeographischer Faktor. *Ann. (Zool.) Soc. Zool. Bot. Fen.* Vanarmo. **10**, 1–262.

Patterson, J. P. (1943). Drosophilidae of the south west. *Univ. Texas Publ. 4313*, 17–216.

Pearce, S. C. (1945). Log normal distributions. *Nature, Lond.* **156**, 747.

Pidgeon, I. M., and Ashby, E. (1940). Studies in applied ecology, I. A statistical analysis of regeneration following protection from grazing. *Proc. Linn. Soc. N.S.W.* **65**, 123–43.

304 PATTERNS IN THE BALANCE OF NATURE

Preston, F. W. (1948). The commonness and rarity of species. *Ecology* **29**, 254–83.
Preston, F. W. (1957). Analysis of Maryland state-wide bird counts. *Maryland Bird Life* **13**, 63–65.
Preston, F. W. (1958). Analysis of the Audubon Christmas counts in terms of the log-normal curve. *Ecology* **39**, 620–4.
Preston, F. W. (1962). The canonical distribution of commonness and rarity. Parts 1 and 2. *Ecology* **43**, 185–215 and 410–32.
Quenouille, M. H. (1949). A relation between the logarithmic, the poisson and the negative binomial series. *Biometrics* **5**, 162–65.
Raunkaier, C. (1934). "Life Forms and Statistical Plant Geography", Oxford University Press, London.
Robert, A. (1955). Les associations de Gyrins dans les étangs et les lacs du Parc du Mont Tremblant. *Canad. Ent.* **87**, 67–78.
Rower, J. A. (1942). Mosquito light trap catches from ten American cities, 1940. *Iowa St. Coll. J. Sci.* **16**, 487–518.
de Sacy, Silvestre (1827). "Chresomathie Arabie", 2nd Edn.
Saunders, A. A. (1936). Ecology of the birds of Quaker Run Valley, Allegany State Park, New York. *N.Y. Stat. Mus. Hdbk.* **16**.
Sharpe, R. B. (1899–1909). "Handlist of the Genera and Species of Birds", 5 vols., London.
Shrieve, F., and Hinckley, A. L. (1937). Thirty years of change in desert vegetation. *Ecology* **18**, 463.
Simmons, F. J. (1943). Occurrence of superparasitism in *Nemeritis canescens*. *Rev. Canad. Biol.* **2**, 15–40.
Simpson, E. H. (1949). Measurement of diversity. *Nature, Lond.* **163**, 688.
Siromoney, C. (1962). Entropy of the logarithmic series distribution. Sankhya, *Indian J. Statist.* A. **24**, 419–20.
Skellam, J. G. (1951). *Biometrics* **7**, 121.
Small, J. (1937–50). Quantitative evolution. A series of papers in *Proc. roy. Soc. Edinb.* and *Proc. R. Irish Acad.*
Spiller, D. (1948). Truncated log-normal and root-normal frequency distributions of insect populations. *Nature, Lond.* **162**, 530.
Sverdrup, H. U., Johnson, M. W., and Fleming, R. H. (1942). "The Oceans", New York. p. 804.
Szidat, L. (1931). *Cordulia aenea* L. ein neuer Hilfswirt für *Prosthogonimus pellucidus* v. Linstow, der Erregen der Trematodenkrankheit der Lebehuhnen. *Zbl. Bakt.* **1** (orig. **119**), 289–93.

Tarshis, L. B. (1958). New data on the biology of *Stilbometopa impressa* Bigot and *Lynchia hirsuta*, Ferris (Diptera, Hippoboscidae). *Ann. ent. Soc Amer.* **51**, 91–105.
Taylor, L. R., and Carter, C. I. (1961). The analysis of numbers and distribution in an aerial population of macro-lepidoptera. *Trans. R. Ent. Soc., Lond.* **113**, 369–86.
Thompson, H. R. (1950). Truncated normal distributions. *Nature, Lond.* **165**, 444.
Tillyard, R. J. (1917). "The Biology of Dragonflies", Cambridge University Press, London.
Tucker, D. G. (1948). Some simple quantitative relationships in ecology with particular reference to birds. *London Nat. 1947*, 42–55.
Tucker, D. G. (1951). The application of the logarithmic series and the index of diversity to bird population studies. *London Nat.* **62**–80.
Uvarov, B., and Johnston, H. B. (1957). A census of the African Acridoid fauna. *Bull. Inst. franç. Afr. noire* A **19**, 511–19.

Vaughan, R. E., and Wiehe, P. O. (1941). Studies in the vegetation of Mauritius. III. *J. Ecol.* **29**, 127–60.

Wadley, F. M. (1950). Notes on the form of distribution of insect and plant populations. *Ann. ent. Soc. Amer.* **43**, 581–6.

Wadley, F. M. (1957). Some mathematical aspects of insect dispersion. *Ann. ent. Soc. Amer.* **50**, 230–1.

Wallace, A. R. (1910). "The World of Life", London.

Waterhouse, G. A. (1932). "What Butterfly is That?" Sydney.

Watson, H. C. (1859). "Cybele Britannica."

Waugh, G. D. (1954). The occurrence of *Mytilicola intestinalis* (Steuer) on the east coast of England. *J. Anim. Ecol.* **23**, 364–7.

Williams, C. B. (1927). A study of butterfly migration in South India and Ceylon. *Trans. Ent. Soc. Lond.* **75**, 1–33.

Williams, C. B. (1930). "The Migration of Butterflies", Oliver and Boyd, Edinburgh.

Williams, C. B. (1937). The use of logarithms in the interpretation of certain entomological problems. *Ann. appl. Biol.* **24**, 404–14.

Williams, C. B. (1940). A note on the statistical analysis of sentence length as a criterion of literary style. *Biometrika* **31**, 356–61.

Williams, C. B. (1943 *a*). See Fisher, Corbet, and Williams (1943).

Williams, C. B. (1943 *b*). Area and number of species. *Nature, Lond.* **152**, 264–5.

Williams, C. B. (1944). Some applications of the logarithmic series and the index of diversity to ecological problems. *J. Ecol.* **32**, 1–44.

Williams, C. B. (1945 *a*). The index of diversity as applied to ecological problems. *Nature, Lond.* **155**, 390–1.

Williams, C. B. (1945 *b*). Recent light-trap catches of Lepidoptera in the U.S.A., analysed in relation to the logarithmic series and the index of diversity. *Ann. ent. Soc. Amer.* **38**, 357–64.

Williams, C. B. (1946). Yule's characteristic and the index of diversity. *Nature, Lond.* **157**, 482.

Williams, C. B. (1947 *a*). The logarithmic series and the comparison of island floras. *Proc. Linn. Soc. Lond.* **158**, 104–8.

Williams, C. B. (1947 *b*). The generic relations of species in small ecological communities. *J. Anim. Ecol.* **16**, 11–18.

Williams, C. B. (1947 *c*). The logarithmic series and its application to biological problems. *J. Ecol.* **34**, 253–72.

Williams, C. B. (1949). Jaccard's generic coefficient and coefficient of floral community in relation to the logarithmic series. *Ann. Bot., Lond.* N.S. **13** (49), 53–58.

Williams, C. B. (1950). The application of the logarithmic series to the frequency of occurrence of plant species in quadrats. *J. Ecol.* **38**, 107–38.

Williams, C. B. (1951 *a*). A note on the relative sizes of genera in the classification of animals and plants. *Proc. Linn. Soc. Lond.* **162**, 171–5.

Williams, C. B. (1951 *b*). Diversity as a measurable character of an animal or plant population. *Année biol.* **27**, 129–41.

Williams, C. B. (1951 *c*). Intra-generic competition as illustrated by Moreau's records of East African birds. *J. Anim. Ecol.* **20**, 246–53.

Williams, C. B. (1952). Sequences of wet and dry days considered in relation to the logarithmic series. *Quart. J. R. Met. Soc.* **78**, 91–96.

Williams, C. B. (1953 *a*). The relative abundance of different species in a wild animal population. *J. Anim. Ecol.* **22**, 14–31.

Williams, C. B. (1953 *b*). Comment on query 96. *Biometrics* **9**, 425–7.

306   PATTERNS IN THE BALANCE OF NATURE

Williams, C. B. (1954). Notes on a small collection of Sphingidae from Nigeria. *Nigerian Field* **19**, 176–9.
Williams, C. B. (1954 a). The statistical outlook in relation to Ecology. *J. Ecology* **42**, 1–13.
Williams, C. B. (1960). The range and pattern of insect abundance. *Amer. Nat.* **94**, 137–51.
Willis, J. C. (1922). "Age and Area", Cambridge University Press, London.
Yule, G. U. (1924). A mathematical theory of evolution based on the conclusions of Dr J. C. Willis. *Phil. Trans.* B. **213**, 21–87.
Yule, G. U. (1944). "The Statistical Study of Literary Vocabulary", Cambridge University Press, London.

APPENDIX A

Some Notes on the Logarithmic Series

Most of the frequency distributions discussed in the previous chapters are fully dealt with in numerous textbooks on statistics. The logarithmic series—first suggested as applicable to population frequency distributions by Fisher (Fisher, Corbet, and Williams, 1943)—is not so frequently dealt with, so a brief account of its properties and methods of fitting is given here.

For fuller details Fisher (in Fisher, Corbet, and Williams, 1943) Williams (1947) and Anscombe (1950) should be consulted.

The Logarithmic Series

The logarithmic series is derived mathematically from the expansion of $\log_e (1 + x)$ and, to avoid negative numbers in the series, may be written as

$$- \log_e (1 - x) = x + x^2/2 + x^3/3 + x^4/4 + x^5/5 \ldots$$

As there are no logs of negative numbers, it follows that x must be less than unity.

When we use the series in biology to express the frequency distribution of groups containing different numbers of units (for example the numbers of species with different numbers of individuals) it may be written

$$n_1, \quad n_1 x/2, \quad n_1 x/3, \quad n_1 x/4, \text{ etc.,}$$

the successive terms being the number of groups with 1, 2, 3, etc., units. No zero term is included.

The series has in this form two parameters, "n_1" which is the number of groups with one unit, and "x" which is a number less than unity. The series has an infinite number of terms, is discontinuous (with only integer values of units per group) and is convergent. The sum of all the groups to infinity is

$$S = \frac{n_1}{x} (- \log_e \overline{1 - x}).$$

The corresponding series of units, i.e. units in all groups in the same abundance class, is

$$n_1, \quad n_1 x, \quad n_1 x^2, \quad n_1 x^3, \quad n_1 x^4, \text{ etc.}$$

This is a geometric series with a constant multiple x. As x is less than unity the series is convergent, and the sum to infinity (N) or the total number of units, equals

$$N = n_1 (1 - x) \text{ or } n_1 = N/(1 - x).$$

From the above it follows that the ratio of the number of groups to the number of units

$$\frac{S}{N} = \frac{1-x}{x} (- \log_e \overline{1-x}).$$

Thus for any particular average number of units per group (the reciprocal of the above) there is only one value of x, and from this n_1 can be obtained. Thus if N and S are known the series is fixed.

If samples by units are taken from a population arranged in a log-series distribution, then the smaller the sample the smaller the average number of units per group and the smaller the value of x: the larger the sample, the greater the average units per group and the greater the value of x. But in all random samples of any size from one population the ratio of n_1 to x is a constant which we have called the "Index of Diversity" or α.

Thus $n_1/x = \alpha$, or $n_1 = \alpha x$.

Since, with increasing size of sample, the value of x increases and approaches unity, it follows that with the same increase in sample size n_1, the number of groups with one unit, increases and gradually approaches the value of α, but can never (theoretically) exceed this.

Thus the log-series distribution can also be written

$$\alpha x, \quad \alpha x^2/2, \quad \alpha x^3/3, \quad \alpha x^4/4, \text{ etc.,}$$

and the sum of this, the total number of groups

$$S = \alpha \log_e (1-x).$$

The index of diversity, α, is therefore a property of the population sampled, but x is a property of the particular sample and depends on its size.

TABLE 146. *The relation between values of "x" and the average number of units per group (N/S) in samples from populations distributed according to the logarithmic series*

x	N/S	x	N/S	x	N/S
0.50	1.443	0.97	9.214	0.9990	144.6
0.60	1.637	0.980	12.53	0.9992	175.1
0.70	1.938	0.985	15.63	0.9994	224.5
0.80	2.483	0.990	21.47	0.9996	319.4
0.85	2.987	0.991	23.38	0.9998	586.9
0.90	3.909	0.992	25.68	0.99990	1086.0
0.91	4.198	0.993	28.58	0.99995	2020
0.92	4.551	0.994	32.38	0.999990	8696
0.93	4.995	0.995	37.48	0.999995	16.390
0.94	5.567	0.996	45.11	0.9999990	71.430
0.95	6.340	0.997	57.21	—	—
0.96	7.458	0.998	80.33	—	—

Since for each value of N/S there is only one value of x, Table 146, and from it Fig. 125, have been prepared to give the approximate value of x for average units per group from 2.5 to 10,000.

Since all samples from a population have the same diversity as the population, it is possible to calculate the increase in number of groups represented as the size of random samples of units is increased. Figure 126 shows the triple relation between units, groups and diversities in the range up to 10,000 units, 330 groups and diversities up to 200. From it we see, for example, that a random sample of 5000 units (e.g. individuals) from a population with a diversity of 20 would be expected to contain 110 groups (e.g. species). If, however, the sample contained only 1000 units, then only seventy-nine groups would be represented.

In extremely large samples with a high average number of units per group x approaches very close to unity. When, for example (see Table 146), the average number of units per group is 145 then $x = 0.9990$. With such a value of x the early terms of the log series approach very closely to the hyperbolic series from which it would be practically indistinguishable in the first few terms.

To fit a log series to a given number of groups (S) and units (N) there are several methods at different levels of accuracy. A very simple method is to use Fig. 126 (within its limits) to find the value of α for the particular combination of N and S. From this n_1, the first term of the series, can be obtained by the formula $n_1 = N\alpha/(N + \alpha)$. Also $x = N/(N + \alpha)$.

Another approximation can be obtained by finding in Fig. 125 the value of x corresponding to the observed value of N/S, the average number of units per group. From this n_1 can be obtained by $n_1 = N(1 - x)$.

Once an approximate value of x has been found, more accurate determinations can be made by successive approximations using the formula

$$\frac{S}{N} = \frac{1-x}{x}(-\log_e \overline{1-x}).$$

Little, however, is gained by excessive accuracy in the determination of x when the biological data are not accurate enough to justify this.

Some of the more useful relations between the number of groups (S), the number of units (N), the number of groups with one unit (n_1), and the diversity (α) are as follows:

(1) $\alpha\ = n_1/x$;

(2) $S\ = \alpha(-\log_e \overline{1-x})$;

(3) $x\ = N(N + \alpha)$;

(4) $\alpha\ = N(1-x)/x$;

(5) $n_1 = N\alpha/(N + \alpha)$;

(6) $n_1 = N(1-x)$;

(7) $N/S = x/(1-x)(-\log_e \overline{1-x})$;

(8) $S\ = \alpha\log_e(1 + N/\alpha)$

or when N/S is large compared with "1", this becomes

(9) $S = \alpha \log_e (N/\alpha)$;

(10) the sum of the 1st moments $= N$

(11) the sum of the 2nd moments $= N (1 - x)^2$.

Formulae for the determination of the median group, and for the summation of the groups up to high levels are given by Gower (1961).

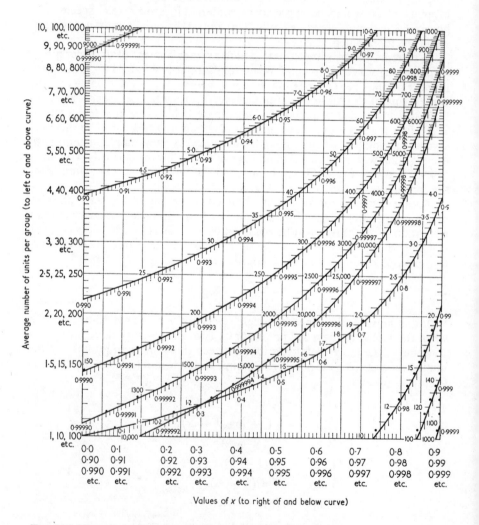

Values of x (to right of and below curve)

FIG. 125. The relation between the average number of units per group (N/S) and the value of "x" in a logarithmic distribution.

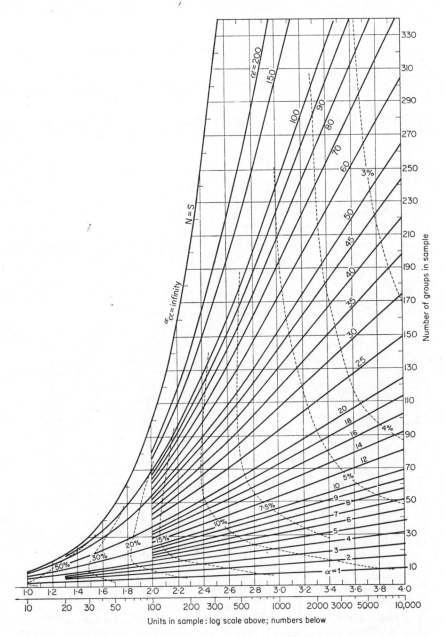

Fig. 126. The relation between the number of groups (*S*), the number of units (*N*), and the diversity (α) in samples of different sizes from populations arranged in a logarithmic series.

APPENDIX B

ORIGINAL DATA FOR THE SPECIES - AREA RELATION OF FLOWERING PLANTS IN FIG. 38 (p. 94)

Contractions
Column 2.
 in = square inch = 6.45 sq cm
 ft = square foot = 144 sq in = 929 sq cm
 yd = square yard = 9 sq ft = 0.84 sq m
 ac = acre = 4840 sq yds = 0.41 hectares
 mi = square mile = 640 ac = 2.59 sq kilometres
 cm = sq centimetre   me = sq metre   km = sq kilometre   mill = million
Column 3.
 If decimal points are given, the values are averages of several plots
Column 4.
 References given at the end of this Appendix
"W." = Wallace, 1910, followed by page reference
"B." = Blake and Atwood, 1942, followed by page reference.

	Locality	Area	No. of Species	Reference
1	Channel Is., Jersey	45 mi	766	Marquand, 1901
2	Channel Is., Guernsey	24.5 mi	804	Marquand, 1901
3	Channel Is., Alderney	3.1 mi	519	Marquand, 1901
4	Channel Is., Sark	2 mi	425	Marquand, 1901
5	Channel Is., Herm	320 ac	255	Marquand, 1901
6	Channel Is., Jathou	44 ac	186	Marquand, 1901
7	Channel Is., Lithou	38 ac	99	Marquand, 1901
8	Channel Is., Crevichou	3 ac	45	Marquand, 1901
9	Channel Is., Burhou	74 ac	18	Marquand, 1901
10 A	England, Jealotts Hill, Grass	2 in	6.9	Blackman, 1935
B	England, Jealotts Hill, Grass	128 in	18.2	Blackman, 1935
11	C. America and Mexico	910,000 mi	12,000	Hemsley, W.42
12	Ireland, Achill Is.	57 mi	414	Praeger
13	Ireland, Clare Is.	6.2	393	Praeger
14	Ireland, Whole Island	32,524 mi	c. 1000	Praeger
15	Great Britain	88,226 mi	c. 1400	
16	Faroe Islands	480 mi	227	Druce
17	Cyprus	3681 mi	1289	Helmbee
18	France	207,000 mi	4260	Costa, 1906
19	Switzerland	16,000 mi	2454	Schinz, 1908
20	Whole World	50 mill mi	170,000	Rendle
21	Ceylon	25,000	2809	Hooker
22	New Zealand	104,000	1763	Cheeseman, 1925
23	Madagascar	288,000 mi	5000	Palacky, 1906
24 A	U.S.A., Michigan, Aspen Ass.	1 me	4.38	Gleason, 1922
B	U.S.A., Michigan, Aspen Ass.	240 me	27	Gleason, 1922
C	U.S.A., Michigan, Aspen Ass.	56 km	80	Gleason, 1922
25	Denmark	29,500 mi	1084	Raunkaier

	Locality	Area	No. of Species	Reference
26	West Indies, St Thomas and St John Is.	32 + 21 mi	904	Raunkaier
27 A	U.S.A., Michigan, Aspen No. 4	1 me	4.2	Gleason, 1925
B	U.S.A., Michigan, Aspen No. 4	100 me	44	Gleason, 1925
C	U.S.A., Michigan, Aspen No. 4	8492 me	84	Gleason, 1925
28 A	U.S.A., Michigan, Aspen No. 5	1 me	4.6	Gleason, 1925
B	U.S.A., Michigan, Aspen No. 5	100 me	34	Gleason, 1925
C	U.S.A., Michigan, Aspen No. 5	4350 me	59	Gleason, 1925
29	Trinidad	1750 mi	2000+	Hart, W.42
30	Jamaica	4200 mi	2702	Brittain, W.42
31	Galapagos	2400 mi	445	Wallace, 63
32	Brasil, Lagoa Santé	66 mi	2490	Warming, W.63
33	Malacca	660 mi	200	Gamble, W.71
34 A	U.S.A., Michigan, Aspen No. 1	1 me	4.8	Gleason, 1925
B	U.S.A., Michigan, Aspen No. 1	4 me	10.7	Gleason, 1925
C	U.S.A., Michigan, Aspen No. 1	16 me	17.4	Gleason, 1925
D	U.S.A., Michigan, Aspen No. 1	64 me	25	Gleason, 1925
E	U.S.A., Michigan, Aspen No. 1	256 me	32	Gleason, 1925
35 A	Java, Tectona Forest	100 me	40	Arrhenius, 1923
B	Java, Tectona Forest	1200 me	104	Arrhenius, 1923
36 A	Java, Tectona Forest	100 me	23	Arrhenius, 1923
B	Java, Tectona Forest	1200 me	64	Arrhenius, 1923
37 A	Java, Tectona Forest	100 me	18.6	Arrhenius, 1923
B	Java, Tectona Forest	1200 me	47	Arrhenius, 1923
38	Singapore	206 mi	1740	Ridley, W.71
39	Penang	107 mi	1813	Curtis, W.71
40 A	Java, Tectona Forest	100 me	5.7	Arrhenius, 1923
B	Java, Tectona Forest	1000 me	8.2	Arrhenius, 1923
41 A	Java, Tectona Forest	100 me	4.5	Arrhenius, 1923
B	Java, Tectona Forest	1000 me	16.7	Arrhenius, 1923
42	U.S.A., Colorado	104,000 mi	2912	Rydberg, 1906
43	England, Buckinghamshire	750 mi	1027	Druce, 1926
44	England, Norfolk	2000 mi	1029	Nicholson, 1914
45	England, Hertfordshire	633 mi	889	Pryor, 1887
46	Java, Panerango (very rich!)	3 km	1750	Koorders, W.74
47	India, Bombay State	111,500 mi	2650	Cooke, 1903
48	England, Scolt Head Is.	1.5 mi	185	Peal, 1934
49	England, Wicken, Cambridge	500 ac	185	Evans, 1923
50	C. and N.E. U.S.A. and adjacent Canada	1 mill mi	3300	Gray, 1908
51	Australia, Queensland	670,500 mi	4700	Bailey, 1909
52	Europe	3.75 mill mi	11,500	Nyman, 1878
53	England, Surrey	700 mi	1081	Beeby, 1902
54	England, Northampton, quarry	50 ft	40	Hepburn, 1942
55	S. Africa, Frankenveldt, purple veldt	1 me	25	West, 1938
56	Balkan Peninsula	187,800 km	6530	Turrill, 1929
57	Arctic, Jan Meynan Is.	304 mi	58	Russell, 1940
58	Arctic, Akpatok Is.	213 mi	123	Polunin, 1934
59	England, Chiltern beechwood	1 me	9	Watt, 1934
60	England, Chiltern beechwood	1 me	10	Watt, 1934
61	British Guiana, Mora forest	3.7 ac	104+	Davis, 1934
62	British Guiana, Mora forest	3.7 ac	143+	Davis, 1934
63 A	S. Africa, Table Mt., sandstone	1 me	11	Adamson, 1931
B	S. Africa, Table Mt., sandstone	50 me	96	Adamson, 1931
64 A	S. Africa, Table Mt., plateau	1 me	14	Adamson, 1931
B	S. Africa, Table Mt., plateau	25 me	44	Adamson, 1931
65	S. Africa, Table Mt., granite soil	25 me	86	Adamson, 1931
66	England, Hampshire woodland	680 ac	121	Adamson, 1921
67	England, Hertford woodland	2–3000 ac	269	Salisbury, 1918

	Locality	Area	No. of Species	Reference
68	Seychelle Is.	150 mi	480	Summerhays, B.100
69	England, fox-covert	10 ac	75	Woodruffe-Peacock, 1918
70 A	Sweden, Stockholm, Herb. Pinus wood	100 cm	4.8	Arrhenius, 1921
B	Sweden, Stockholm, Herb. Pinus wood	400 cm	9.8	Arrhenius, 1921
C	Sweden, Stockholm, Herb. Pinus wood	3200 cm	23	Arrhenius, 1921
D	Sweden, Stockholm, Herb. Pinus wood	1 me	33	Arrhenius, 1921
71 A	Sweden, Stockholm, Herb. Picea wood	100 cm	2.5	Arrhenius, 1921
B	Sweden, Stockholm, Herb. Picea wood	400 cm	5.4	Arrhenius, 1921
C	Sweden, Stockholm, Herb. Picea wood	1600 cm	10.2	Arrhenius, 1921
D	Sweden, Stockholm, Herb. Picea wood	6400 cm	16.5	Arrhenius, 1921
72	Java, Kambangen Is. (very rich)	3 km	2400	Koorders, W.71
73 A	Sweden, Stockholm, Vaccinium Vitis, Pinus Assoc.	100 cm	1.4	Arrhenius, 1921
B	Sweden, Stockholm, Vaccinium Vitis, Pinus Assoc.	400 cm	2.0	Arrhenius, 1921
C	Sweden, Stockholm, Vaccinium Vitis, Pinus Assoc.	3200 cm	2.3	Arrhenius, 1921
D	Sweden, Stockholm, Vaccinium Vitis, Pinus Assoc.	1 me	3.0	Arrhenius, 1921
74	Australia, Victoria, desert	c. 500 mi	225	D'Alton, 1913, B.74
75	Aden	15 mi	250	Blatter, 1914
76	England, Herts, "Broadbalk Wilderness"	0.5 ac	79	Brenchley, 1915
77	England, Herts, "Geescroft"	2.3 ac	88	Brenchley, 1915
78	U.S.A., Connecticut	4965 mi	2228	Graves, B. 171
79	Falkland Is., S. Atlantic	6500 mi	162	Skottsberg, 1913
80	England, Wye Gorge (very rich)	3.2 km	700	Armitage, 1914
81	Panama, Barra Colorado Is.	14.6 km	715	Kenoyer, 1929
82 A	U.S.A., Florida, scrub	10,000 ft	20	Mulviana, 1931
B	U.S.A., Florida, scrub	45,000 ft	32	Mulviana, 1931
83	U.S.A., Michigan, hardwood forest	2500 me	79	Cain, 1935
84	U.S.A., Pennsylvania, forest	53.1 ac	73	Hough, 1936
85	England, Herts, wheatfield weeds, plot 8	0.5 ac	20	Warington, 1924
86	England, Herts, wheatfield weeds, plot 3	0.5 ac	30	Warington, 1924
87	U.S.A., Wisconsin, maple-basswood	2100 me	58	Eggler, 1938
88	U.S.A., Arizona, Tucson, desert	10,000 me	6	Shreve, 1937
89	U.S.A., Arizona, Tucson, desert	10,000 me	18	Shreve, 1937
90	U.S.A., Minnesota, sand-bar	29.7 ac	119	Laleka, 1939
91	U.S.A., Missouri, Horn Is.	6–7 mi	51	Pessin, 1941
92	West Indies, St Croix, Sandy Point	1–2 km	80	Raunkaier
93	Australia, Sydney, Parametta river	20 mi	618	Deane, W.38
94	Italy, Capri	4 mi	719	Beguinot, W.71
95	England, Kent	12 ft	20	Darwin, W.27
96 A	Denmark, Oakwood	0.1 me	3	Raunkaier
B	Denmark, Oakwood	2.5 m	12	Raunkaier
97	England, Cheshire, old quarry	2 ac	170	in litt.

	Locality	Area	No. of Species	Reference
98	Maldive Is.	115 mi	156	Bond, in litt.
99	Bahrain Is. (Persian Gulf)	c. 150 mi	150+	Good, in litt.
100	West Indies, Martinique	400 mi	1798	Stehle, B. 229
101	West Indies, Trinidad	1750 mi	2200	Beard
102	Venezuela,	942,000 km	12,000	Pittier, B. 260
103	Juan Fernández, Pacific, off Chile	141 km	143	Skottsberg, B. 114
104	France, South, sand-dunes	5 me	10	Raunkaier
105	Baffins Land, Arctic Canada	236,000 mi	129	Raunkaier
106	Canada, Labrador, Chidley Penin.	c. 200 mi	64	Raunkaier
107	Samos Is., Mediterranean	190 mi	400	Raunkaier
108	Arctic, Spitzbergen	24,000 mi	110	Raunkaier
109	Iceland	40,500 mi	329	Raunkaier
110	Arctic, Novaya Zembla	35,000 mi	192	Raunkaier
111	Kerguelen Is., S. Indian Ocean	2500 mi	21	Raunkaier
112	Tristan Da Cunha, S. Atlantic	45 mi	29	Raunkaier
113	Italy, Argentaria	(40 mi)	866	Raunkaier
114	Madeira Is., Atlantic	315 mi	427	Raunkaier
115	Balearic Is., Mediterranean	1860 mi	1100	Jaccard, 1939
116	Formosa	13,000 mi	1613	Raunkaier
117	Sardinia	9200 mi	1793	Raunkaier
118	Greece, Attica	2481 mi	1092	Raunkaier
119	Aegean Is., Mediterranean	32 mi	566	Raunkaier
120	Karpathos Is., Mediterranean	150 mi	375	Raunkaier
121	Zante Is., Mediterranean	165 mi	626	Raunkaier
122	Tremitic and Pelagos Is., Mediterranean	2 mi	448	Raunkaier
123 A	England, Surrey	1 mi	400	Watson, 1859
B	England, Surrey	10 mi	600	Watson, 1859
C	England, Surrey	60 mi	660	Watson, 1859
D	England, Surrey, whole county	760 mi	840	Watson, 1859
E	England, South Thames area	2316 mi	972	Watson, 1859
F	England, Thames area	7007 mi	1051	Watson, 1859
G	England, Southern area	38,474	1280	Watson, 1859
H	England, whole country	87,417	1425	Watson, 1859
124	Algeria	222,000 mi	3316	Battandier, 1904, B.17
125	Congo, Katanga	180,000 mi	2230	Wildeman, 1927, B.26
126	Somaliland, British	68,000	318	Drake-Brockerman, 1912, B.17
127	S. Africa, White Hill District	40 mi	700	Compton, 1931, B.27
128	S. Africa, Kaffraria	16,000 mi	2449	Sim, 1894, B.29
129	Egypt	363,000 mi	1800	Simpson, 1930, B.30
130	Sinai Peninsula	30,000 mi	942	Zohary, 1935, B.32
131	Gambia, W. Africa	4000 mi	285	Williams, 1907, B.38
132	Natal, S. Africa	35,284 mi	3786	Bews, 1421, B.48
133	S. Africa, Transvaal and Swaziland	117,000 mi	c. 4500	Burt-Davey, 1926, B.64
134	Australia, Brisbane	c. 1300 mi	1228	Bailey, 1879, B.69
135	Australia, extra tropical	c. 500,000 mi	1935	Tate, 1890, B.70
136	Australia, Kangaroo Is.	1679 mi	653	Cleland, 1927, B.71
137	New Zealand, Stewart Is.	6–700 mi	450	Kirk, 1885, B.83
138	Tasmania	26,215 mi	1096	Spicer, 1878, B.86
139	Ascension Is., Atlantic	38 mi	21	Hemsley, 1884, B.86
140	Azores, Atlantic	922 mi	478	Watson, 1870, B.87
141	Canary Is., Atlantic	2807 mi	1352	Pitard, 1908, B.87
142	Cape Verde Is., Atlantic	1516 mi	300	Chevalier, 1935, B.88
143	Madeira Is., Atlantic	.314 mi	951	Menezes, 1914, B.90
144	South Georgia, S. Atlantic	1094 mi	19	Schenk, 1905, B.91
145	Aldabra Is., Indian Ocean	60 mi	171	Hemsley, 1919, B.93

	Locality	Area	No. of Species	Reference
146	Andaman Is., Indian Ocean	2508 mi	540	Parkinson, 1923, B.93
147	Christmas Is., Indian Ocean	60 mi	151	Ridley, 1906, B.94
148	Laccadive Is., Indian Ocean	c. 80 mi	194	Prain, 1893, B.96
149	Mauritius and Seychelles	876 mi	1327	Baker, 1877, B.98
150	Chatham Is., S. Pacific	372 mi	209	Olivier, 1917, B.98
151	Lord Howe Is.	5 mi	217	Maiden, B.102
152	Antipodes Is., S. Pacific	24 mi	55	Kirk, B.102
153	Norfolk Is.	13 mi	175	Laing, B.102
154	Borneo	307,000 mi	4924+	Merrill, B.103
155	Guam Is., Pacific	225 mi	545	Merrill, B.117
156	Philippine Is.	114,000 mi	8120	Merrill, B.122
157	Philippine Is., Manila	100 km	1007	Merrill, B.123
158	Rorotonga, Cook Is., Pacific	31 mi	334	Cheeseman, B.125
159	Tonga and Friendly Is., Pacific	250 mi	290	Hemsley, B.126
160	Tahiti, Pacific	400 mi	417	Nadeau, B.129
161	Alaska	586,000 mi	684	Porsild, B.130
162	Canada	3.7 mill mi	3209	Macoun, B.133
163	Canada, Manitoba Province	246,000 mi	1029	Jackson, B.139
164	Canada, St Pierre and Miguelon Is.	81 mi	538	Waghorn, B.140
165	Brit. Honduras	8600 mi	3000	Lundell,
166	Guatemala	42,500 mi	(3736)	Smith, B.147
167	Salvador	23,180 mi	2070	Standley, B.154
168	Greenland	830,000	390	Ostenfeld, B.155
169	North America, north of Mexico	7.5 mill mi	16,673	Hellar, B.158
170	U.S.A., Arkansas State	53,336 mi	1610+	Branner, B.164
171	U.S.A., California	158,297 mi	3727	Jepson, B.165
172	U.S.A., Dist. Columbia (15 mi radius)	700 mi	1343	Hitchcock, B.172
173	U.S.A., Indiana	36,354 mi	2109	McDonald, B.176
174	U.S.A., Iowa	56,147 mi	1263	Cratty, B.178
175	U.S.A., Kansas	82,158 mi	1933	Smyth, B.181
176	U.S.A., Maryland	12,327 mi	1400	Shrove, B.186
177	U.S.A., Michigan	58,000 mi	2365	Beal, B.189
178	Cuba	44,000 mi	8000	Carabia
179 A	Denmark, sedge meadow	0.1 me	17.6	Raunkaier
B	Denmark, sedge meadow	5.0 me	78	Raunkaier
180	U.S.A., Utah and Nevada	205,000 mi	3700	Tidestrom, B.196
181	U.S.A., New Jersey	8224 mi	1999	Britton, B.197
182	U.S.A., New Mexico, desert	270 mi	62	Emerson, B.199
183	Denmark, field boundaries	0.1 me	12	Raunkaier
184	Denmark, field boundaries	5.0 me	78	Raunkaier
185	U.S.A., Texas	266,000 mi	5099	Cory, B.215
186	Bahamas Is., W. Indies	4400 mi	1021	Britton, B.224
187	Cuba, Isle of Pines	840 mi	731	Jennings, B.226
188	Haiti, Tortue Is.	116 mi	889	Ekman, B.227
189	Jamaica	4207 mi	2412+	Faucet, B.227
190	W. Indies, St Thomas, St Juan and St Croix	130	1052	Britten, B.229
192	Brasil	3.3 mill. mi	22,757	Martius, B.239
193	Chile	290,000 mi	5358	Philippi, B.245
194	San Thomé ⎫	c. 1000 km	556	Exell, 1946
195	Fernando Po ⎬ Atlantic off	c. 2000 km	842	Exell 1946
196	Principe ⎰ W. African Coast	c. 126 km	276	Exell, 1946
197	Annobon ⎭	c. 14 km	115	Exell, 1946
198	Scotland, N. Rona Is.	300 ac	43	Darling, 1947
199	Switzerland, Joux Valley	260 km	827	Aubert (Raunkaier, 1934)
200	Switzerland, Val d'Anniviers	10 me	46	Jaccard, 1939
201	Scotland, St Kilda Is.	2100 ac	140	Darling, 1947
202	Scotland, Raasay and other Is.	44 mi	500	Harrison, 1937

	Locality	Area	No. of Species	Reference
203	British Guiana	90,000 mi	3254	Schomberg, W.21
204	Australia and Tasmania	3 mill. mi	4200	Brown, W.21
206	England, Cadney, Lincolnshire	just over 3 mi	720	Wallace, 1910, p. 26
207	England, Edmondsham, Dorset	under 3 mi	640	Wallace, 1910, p. 26
208	England, Lincoln and Leicester	10–15 ac	50–60	Woodruffe-Peacock, W.26
209	England, Lincoln and Leicester	273 ft	20–30	Woodruffe-Peacock, W.26
210	Lapland	150,000 mi	500	de Candolle, W.29
211	Scandinavia and Denmark	315,000 mi	1677	de Candolle, W.29
212	Sweden	173,000 mi	1165	de Candolle, W.29
213	Germany	208,000 mi	2547	Garche, W.29
214	Sardinia	9300 mi	1770	Beccari, W.29
215	Italy	91,400 mi	4350	Beccari, W.29
216	Sicily	9940 mi	2070	Beccari, W.29
217	U.S.A., S.E. States	630,000 mi	6321	Cockerell, W.30
218	Mediterranean	550,000 mi	7000	Tchikatcheff, W.31
219	E. Europe and S.W. Asia	2 mill mi	11,876	Boissier, W.31
220	China and Korea	1.5 mill mi	8200+	Hemsley, W.33
221	Japan	150,000 mi	4000	Hayati, W.32
222	New South Wales	310,700 mi	3105	Muller, W.32
223	Western Australia	90,000 mi excl. desert	3242	Muller, W.33
224	South Africa	1 mill mi	13,000+	Thomer, W.33
225	South Africa, Cape Region	30,000 mi	4500	Bolus, W.33
226	New Zealand	103,650 mi	1474	Cheeseman, W.33
229	U.S.A., District of Columbia	108 mi	922	Ward, W.33
230	Japan, Mount Nikko	360 mi	800	Hayati, W.36
232	South Africa, Cape Peninsula	197 mi	1750	Bolus, W.33
233	Africa, South of Sahara	6.5 mill mi	18,000	Thomas, W.42
234	Australia, N.S.W., Cumberland Co	1400 mi	1213	Woolls, W.32
235	Malay Peninsula	35,000 mi	5100	Gamble, W.42
236	Burma	172,000 mi	6000	Hooker, W.42
237	Sweden, Härjedal	5375 mi	606	Birger, W. 34
238	England, Malvern Hills	120 mi	802	de Candolle, W.34
239	Italy, Susa, Piedmont	540 mi	2203	Beccari, W.34
240	Switzerland, Poschiavo	92 mi	1200	Field, W.34
241	Switzerland, Schaffhauser	114 mi	1220	Field, W.34
242	Switzerland, Thurgau	381 mi	1006	Field, W.34
243	Switzerland, Grisons	2773 mi	1550	Field, W.34
244	Switzerland, Valais	2027 mi	1752	Field, W.34

ORIGINAL REFERENCE FOR FIG. 38 AND APPENDIX B

Adamson, R. S. (1921). *J.Ecol.* **9**, 114. Adamson, R. S. (1931). *J.Ecol.* **19**, 304. Armitage, E. (1914). *J. Ecol.* **2**, 98. Arrhenius, O. (1921). *J.Ecol.* **9**, 95. Arrhenius, O. (1923). *Ecology* **4**, 68.

Bailey, A. M. (1909). "Catalogue of Queensland Plants", Brisbane. Beeby (1902). "Victoria County Histories", Surrey. Blackman, G. E. (1935). *Ann. Bot.* **49**, 749. Blake, S. F., and Atwood, A. C. (1942). *U.S. Dept Agr. Misc. Pub.* 401. Blatter, E. (1926). *Rec. Bot. Survey India* **7**, 1. Brenchley, W. E., and Adam, K. (1915). *J. Ecol.* **8**, 193.

Cain, S. A. (1935). *Ecology* **16**, 500. Cooke, T. (1903). "Flora of Bombay State", London.

D'Alton (1913). *Vict. Nat.* **30**, 65. Darling, F. F. (1947). "Natural History of the Highlands and Islands", London. Davis, T. W. A., and Richards, P. W. (1934). *J. Ecol.* **22**, 134. Druce, G. C. (1926). "Flora of Buckinghamshire".

Eggler, W. A. (1938). *Ecology* 19, 252. Evans, A. H. (1923). *In* Steer's "Natural History of Wicken Fen", Cambridge University Press, London. Exell, A. W. (1944). "Catalogue. Vascular Plants of S. Tomé", London.

Gleason, H. A. (1922). *Ecology* 3, 158. Gleason, H. A. (1925). *Ecology* 6, 66 Gray, A. (ed.) (1908). "Manual of Botany", 7th Edition. New York.

Harrison, J. W. H. (1937). *Proc. Univ. Durham Phil. Soc.*. 9, 260. Hepburn, J. (1942). *J. Ecol.* 30, 61. Hough, A. F. (1936). *Ecology* 17, 9.

Jaccard, P. (1939). *Bull. Soc. Vaud Sc. Nat.* 60, 249.

Kenoyer, L. A. (1929). *Ecology* 10, 201.

Lakela, O. (1939). *Ecology* 20, 544. Lundell, C. L. (1937). *Carnegie Inst. Washington Pub.* 478.

Marquand, E. D. (1901). "Flora of Guernsey and the Lesser Channel Islands", London. Mulviana, M. (1931). *Ecology* 12, 531.

Nicholson, W. A. (1914). "A Flora of Norfolk." Nyman, C. F. (1878–87). "Conspectus Florae Europae."

Peal (1934). *In* Steer's "Scolt Head Island".

Pessin and Burleigh (1941). *Ecology* 22, 70.

Polunin, N. (1934). *J. Ecol.* 22, 345. Praeger, R. L. (1934). "A Botanist in Ireland." Pryor (1887). "Flora of Hertfordshire."

Raunkaier, C. (1934). "Life Forms." Russell, R. S., and Wellington, P. S. (1940). *J. Ecol.* 28, 159. Rydberg, P. A. (1906). "Flora of Colorado" (Fort Collins, Col.).

Salisbury, E. J. (1918). *J. Ecol.* 6, 14. Shreve, R., and Hinchley, A. L. (1937). *Ecology* 18, 463. Skottsberg, C. (1913). *Sven. Vet. Akad.* 50. Simpson, N. D. (1930). *Tech. Bull. Min. Agr. Egypt*, no. 93.

Turrill, W. B. (1929). "Plant Life of the Balkan Peninsula", Oxford University Press, London.

Wallace, A. R. (1910). "The World of Life", London. Watson, H. C. (1859). *Cybele Britannica* 4, 379–81. Watt, A. S. (1934). *J. Ecol.* 22, 246. West, O. (1938). *J. Ecol.* 26, 212. Woodruffe-Peacock, E. A. (1918). *J. Ecol.* 6, 110.

APPENDIX C

Species	Total 4 years	Years ex 4	Species	Total 4 years	Years ex 4
1 Common Loon	3	3	37 Red-breast		
2 Red-throated Loon	1	1	Merganser	4633	4
3 Red-necked Grebe	1	1	38 Goshawk	5	3
4 Horned Grebe	138	4	39 Sharp-shinned Hawk	21	4
5 Eared Grebe	1	1	40 Cooper's Hawk	52	4
6 Western Grebe	1	1	41 Red-tailed Hawk	456	4
7 Pied-billed Grebe	14	4	42 Krider's Red T.		
8 Double-crest			Hawk	3	2
Cormorant	4	3	43 Red-shouldered		
9 Great Blue Heron	25	4	Hawk	150	4
10 Black-crown Night			44 Broad-winged Hawk	1	1
Heron	9	2	45 Rough-legged Hawk	92	4
11 Canada Goose	3194	4	46 Ferruginous R.L.		
12 Snow Goose	1	1	Hawk	1	1
13 Blue Goose	3	1	47 Bald Eagle	254	4
14 Hutchin's Goose	2	1	48 Marsh Hawk	168	4
15 Mallard	804,806	4	49 Osprey	2	1
16 Black Duck	20,629	4	50 Peregrine Falcon	5	1
17 Gadwell	10	3	51 Pigeon Hawk	4	3
18 Baldpate (Wigeon)	55	4	52 Sparrow Hawk	354	4
19 Pintail	67	4	53 Prairie Chicken	38	1
20 Green-winged Teal	26	3	54 Bobwhite Quail	900	4
21 Shoveller	5	1	55 Ring-necked		
22 Wood Duck	2	1	Pheasant	590	4
23 Red Head	69	4	56 Grey Partridge	26	1
24 Ring-necked Dove	57	4	57 American Coot	1936	4
25 Canvas Duck	3776	4	58 Kill Deer	47	4
26 Greater Scaup Duck	94	4	59 Wilson's Snipe	11	3
27 Lesser Scaup	3711	4	60 Glaucous Gull	1	1
28 American Golden Eye	8777	4	61 Iceland Gull	1	1
29 Barrow's Golden Eye	1	1	62 Herring Gull	18,630	4
30 Buffle Head	666	4	63 Ring-billed Gull	8237	4
31 Old Squaw	7746	4	64 Franklin's Gull	32	1
32 White-winged Scoter	96	4	65 Bonapart's Gull	1688	3
33 Surf Scoter	4	3	66 Little Gull	4	3
34 Ruddy Duck	86	4	67 Mourning Dove	1547	4
35 Hooded Merganser	58	4	68 Barn Owl	3	3
36 Common (Am.)			69 Screech Owl	46	4
Merganser	4884	4	70 Great Horned Owl	44	4

Species	Total 4 years	Years ex 4	Species	Total 4 years	Years ex 4
71 Snowy Owl	2	2	109 Common Starling	95,408	4
72 Barred Owl	67	4	110 Myrtle Warbler	8	1
73 Long-eared Owl	130	4	111 Northern Water		
74 Short-eared Owl	20	3	Thrush	1	1
75 Saw-whet Owl	7	4	112 House Sparrow	36,344	4
76 Belted Kingfisher	89	4	113 Eastern Meadow		
77 Yellow-shafted			Lark	1052	4
Flicker	473	4	114 Western Meadow		
78 Pileated Woodpecker	20	4	Lark	16	3
79 Red-bellied Wood-			115 Red-wing Black		
pecker	980	4	Bird	8608	4
80 Red-headed Wood-			116 Rusty Blackbird	853	3
pecker	541	4	117 Brewer's Blackbird	2	2
81 Yellow-bellied			118 Common Purple		
Sap-sucker	16	4	Crackle	1052	4
82 Hairy Woodpecker	530	4	119 Brown-head Cow-		
83 Downy Woodpecker	2105	4	Bird	553	4
84 Eastern Phoebe	1	1	120 Cardinal	4868	4
85 Northern Horned			121 Evening Grosbeck	153	3
Lark	6	1	122 Purple Finch	453	4
86 Prairie Horned Lark	1825	4	123 Pine Grosbeck	20	3
87 Blue Jay	3944	4	124 Common Redpoll	153	4
88 American Crow	28,487	4	125 Pine Siskin	440	4
89 Black-capped			126 Common Amer.		
Chickadee	4450	4	Goldfinch	3332	4
90 Carolina Chickadee	280	4	127 Red Crossbill	18	3
91 Tufted Titmouse	2374	4	128 White-wing		
92 White-breasted			Crossbill	72	2
Nuthatch	1347	4	129 Eastern Towher	36	4
93 Red-breasted			130 Savannah Sparrow	1	1
Nuthatch	168	4	131 Vesper Sparrow	3	2
94 Brown Creeper	235	4	132 Slate-coloured		
95 Winter Wren	35	4	Junco	14,554	4
96 Carolina Wren	368	4	133 Oregon Junco	22	4
97 Mocking Bird	274	4	134 Pink-sided Junco	1	1
98 Catbird	3	2	135 Tree Sparrow	16,775	4
99 Brown Thrasher	8	3	136 Chipping Sparrow	38	2
100 Robin	767	4	137 Field Sparrow	149	4
101 Hermit Thrush	4	3	138 Harris's Sparrow	3	1
102 Varied Thrush	2	1	139 White-crowned		
103 Eastern Bluebird	315	4	Sparrow	182	4
104 Gold-crowned			140 White-throated		
Ringlet	410	4	Sparrow	83	4
105 Ruby-crowned			141 Fox's Sparrow	5	3
Ringlet	15	3	142 Lincoln's Sparrow	1	1
106 Cedar Waxwing	1848	4	143 Swamp Sparrow	138	4
107 Northern Grey			144 Song Sparrow	1113	4
Shrike	22	4	145 Lapland Longspur	516	4
108 Loggerhead Shrike	56	4	146 Snow Bunting	66	3

INDEX